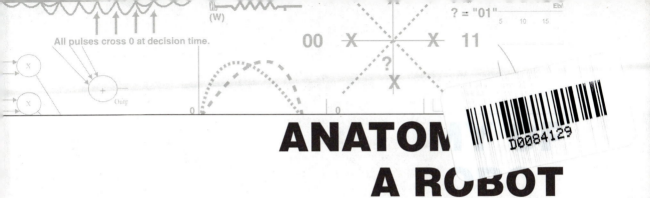

ANATOM A ROBOT

CHARLES M. BERGREN

McGraw-Hill

New York   Chicago   San Francisco   Lisbon   London   Madrid
Mexico City   Milan   New Delhi   San Juan   Seoul
Singapore   Sydney   Toronto

The McGraw·Hill Companies

Library of Congress Cataloging-in-Publication Data

Bergren, Charles M.
   Anatomy of a robot / Charles M. Bergren
     p. cm.
   Includes bibliographical references and index.
   ISBN 0-07-141657-9 (alk. paper)
    I. Robots--Design and contruction.   II. Title

   TJ211.B485 2003
   629.8'92—dc21   2003046335

1 2 3 4 5 6 7 8 9 0 DOC/DOC 0 9 8 7 6 5 4 3

ISBN 0-07-141657-9

The sponsoring editor for this book was Judy Bass and the production supervisor was Pamela A. Pelton. It was set in Century Schoolbook by MacAllister Publishing Services, LLC.

Printed and bound by RR Donnelley.

This book is printed on recycled, acid-free paper containing a minimum of 50 percent recycled de-inked fiber.

McGraw-Hill books are available at special quantity discounts to use as premiums and sales promotions, or for use in corporate training programs. For more information, please write to the Director of Special Sales, Professional Publishing, McGraw-Hill, Two Penn Plaza, New York, NY 10121-2298. Or contact your local bookstore.

To my son and my wonderful family

CONTENTS

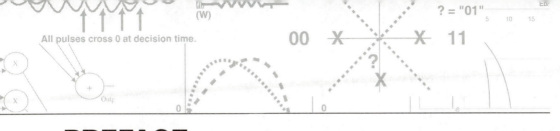

PREFACE

Two years ago, I took my six-year-old son to a "robot race" up in the Rockies near Boulder. It was held in the community center of a small mountain town. Nevertheless, it was packed with about 100 enthusiastic people and many interesting exhibits. The central event was to be a timed race along a prescribed course. Several school-aged kids had entered plastic robots clearly built from parts from the same toy manufacturer. The racecourse was a plastic mat approximately 15 feet on each side. The robots had to follow a one-inch-wide, serpentine black line on the mat from beginning to end. The winner would be the robot finishing with the fastest time.

I watched the kids tuning up their robots on the racecourse before the race. Each robot had a sensor on each side that could detect the black line. If the robot moved forward and started to cross the line, the electronics would correct the steering and move the robot back on course.

It was clear the kids were all having trouble. None of the robots could follow the course from beginning to end. They would invariably lurch too far over the black race-course line and get lost, spinning in useless circles. Legions of adult advisors huddled with the kids, making all sorts of changes, yet nobody was making progress. To me, the answer was obvious and I wanted to help.

Off in the corner, a bit cowed and unsure of himself, was the youngest competitor. Let's call him Sam. He may have been 13 and was there with his mom. They, too, were making changes without good results. I approached Sam's mom, discretely asked permission to help, and joined their team. Without going into the theory, I explained that the robots were all too fast and powerful for their own control systems. I recommended slowing down Sam's robot by adding more weight at the back end. We finally decided to build a sled for the robot to drag and set about finding the materials. With the race deadline approaching, Sam himself came up with the solution. With a quick glance to ask permission, he grabbed his mom's handheld camera and slipped the wrist strap over a post on the rear of the robot. We confirmed the robot could still move slowly down the racecourse line towing the camera. Sam took the batteries out of the camera until it was near the right weight. All too soon, race time came and we had to halt our experiment.

One after another, the older competitors' robots raced down the course only to stray off the black line and be disqualified. A couple of the robots did finish after wandering around lost and wasting a good deal of time. Eventually, the time came for Sam to race his robot. He placed his robot on the starting line, plopped his mom's camera down behind it, confidently put the wrist strap on the rear bumper, and pushed the start button with a bit of ceremony. As Sam's robot left the starting line, it lurched forward, tugging the camera behind it. The crowd started to buzz and I watched the highly amused advisors talking among themselves. It was clear some of them understood what was going on.

To make a long story short, Sam's robot reliably chugged around the racecourse and he won. The look on his face alone was worth the effort. Sam's nominal reward was a kit for a bigger robot, but I think he walked away with much more than that.

After the race, Sam was eager to know how I knew the solution. I took Sam aside and gave him a glimpse of the college-level mathematics and graphs that were behind his victory. My intention was to stimulate his curiosity and point him in the direction that would lead him to further accomplishments. I went home feeling wonderful, proud for myself, and happy for Sam.

After all, everyone seeks direction in life. We experience a feeling of comfort when we discover that our problems are definitive, comprehensible, and tractable. To build a successful robot, it takes a disciplined approach. Many pitfalls are possible, but they are not inevitable. The subjects you will have to master are many and difficult, but not incomprehensible.

To be clear, it is not the intention of my book to teach you how to build a robot. Others can find the nuts and bolts better than I, but if you want to come away enriched with the seminal knowledge of the academic and professional disciplines necessary to be successful in the field, then this book is for you. Each major discipline is the subject of a separate chapter. Each chapter will cover the basics but will also lead you to theory and reasoning that can capture the imagination. For each discipline, legions of engineers and professors spend their entire careers sweating the details.

Sam, if you're out there, I hope one of them is you.

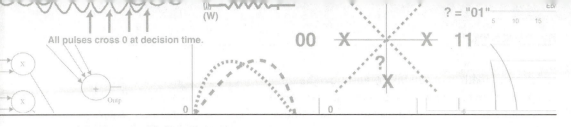

INTRODUCTION

The boundless energy of youth often must give way to the laws of physics. All too often I've seen bright ideas flounder for a lack of fundamental knowledge. If this book can foster the development of the art, if it can encourage and educate the robotic community, if it can provide the missing ingredients—the secret sauce—then I did my job right. If you have a sense that a robot is more than wires and wheels, then this book is for you.

Math rules physics, and physics rules robots. This book sheds light on the math and physics behind a robot design, and does so in an accessible way. The text was written for all ages, from high school through college and beyond. The math used in the book includes algebra, calculus, and differential equations. For readers unacquainted with these subjects, I made sure the text "returns to Earth" often. Nobody should be left behind. The laws of physics and math are evident in everyday life, and several examples are given in this book. Throughout the history of science and technology, the path to great discoveries has almost always started with the observation of simple events. Newton's apple, Einstein's empty room in space, and Shannon's word games are clear examples. Proceeding from an intuitive, personal understanding of the basic laws of physics and math, you can take your understanding further. Using this knowledge, you can predict the behavior of your robot in advance. As problems crop up, you'll have the basic knowledge to move effectively toward solutions.

Throughout the book, I've also thrown in experience gained from 32 years of engineering design. I can't be there when you build your robot, but I can put tools in your belt and pass on such wisdom as we both can sit still for.

Originally, I started this project for the fame, fortune, and groupies. As the chapters rolled out, I got my true rewards. I relearned the basic technologies to better explain them. I dug into the larger questions lurking behind the equations and technology. And as the book developed, I found an outlet for other thoughts I've had for quite some time. I hope my philosophical asides prove entertaining.

The book is divided into chapters that deal with monolithic subjects like computer hardware, computer software, *digital signal processing* (DSP), communications, power, and control systems. It is my hope that readers will find these individual subjects compelling enough to pursue them further. In each chapter, I've included URLs for web sites that explain the technologies in more depth. The Web can be a great place to obtain a continuing education.

Chapter 1 covers project management. More robots bite the dust for a lack of management discipline than any other reason. Building robots is much like going into battle. You can do great damage coming straight out of the gate and swinging swords, but it takes planning to make sure only the enemy gets cut. The chapter outlines how to approach a robot project from the outset. It includes development process flowcharts, checklists, and lots of tips. Robots are not built; they are born. With forethought and preparation, the process can be much less painful. And lest we forget, the project depends on people. Motivation and management, of self and others, are required for success.

Chapter 2 covers control systems. This is a complex field with a language of its own and many disciplines. If someone were to gather data about why robot projects fail, I'm guessing mechanics and power problems would come first. Control system problems would be right up there, too. The chapter discusses control system architecture; distributed and centralized control systems are compared and contrasted. Most robots have centralized systems and use open-loop and closed-loop control methods. The text outlines the basic behavior of a second order-control system, a good model for the behavior of many robotic systems. The text explains the math needed to understand and control system behavior. Specific examples of ways to design and correct such a control system are also given. Last of all, I've thrown in all the tricks of the trade that I know.

Chapter 3 covers computer hardware. I've outlined many of the reasons for using a computer in a robot and ways to accelerate the design process. Several computer architectures are listed, including analog, general-purpose digital, DSP, neural networks, and parallel processors. I've outlined the basic architecture of general-purpose digital microprocessors and commented on the applicability of various computer options. Just as the lack of planning can ruin a robot project, so too can the wrong choice of micro-processor. The last part of the chapter has a large checklist that can help you through the process of selecting a computer.

Chapter 4 covers reliability, safety, and compliance. The first section defines relia-bility and provides methods for predicting and measuring it. The chapter also includes a list of components to be wary of and some advice about using them. In the safety sec-tion is a list of dangers that can sneak up on even the most experienced designers, and it also offers advice about managing risks. The compliance and testing section covers environmental considerations, emissions, and many tips for forestalling problems.

Chapter 5 covers the early stage of the design process, the *high-level design* (HLD). The text covers where to start, what to consider first, and how to make the design gel early. Although every robotic project will be different, I wanted the chapter to document how I would go about designing a robot. I closed my eyes, gave myself a phantom team of engineers, and wrote down what I'd do. Let me know if you'd do it differently.

Chapter 6 covers power and energy. First, I discuss how to determine the robot's energy requirements. It outlines a series of considerations that should be taken into account in the selection and use of an energy source, with a specific concentration on batteries.

Chapter 7 covers energy and software control systems, with an emphasis on energy management. It includes a list of specific actions to take in the design of an energy-efficient robot. I mentioned many considerations that should be kept in mind during the selection and design of robotic software. The chapter outlines a coordinated approach to the selection of a processor, a battery, a power supply, operating software, and appli-cation software. Included are many software techniques that have proven successful, including a discussion of braking methods.

Chapter 8 covers DSP and the chapter starts with an example of DSP processing that is familiar to all of us. This leads to the two basic theorems of DSP. Specific examples illustrate the need for both learning and using the theorems. The chapter includes dif-ferent methods of constructing a classic DSP control system. I've included rules of thumb for picking components, methods for programming them, and ways to test them.

Chapter 9 covers communication, which is vital to the effectiveness and power of people, and robots are not far behind in this need. The chapter starts with the definition of communication, the concept of noise, and Shannon's theorem for the capacity of a noisy communications link. I discuss baseband transmission, the basic techniques for sending pulses down a wire, and the common baseband communication links, includ-ing the Ethernet. The chapter outlines the reasons for modulated communication and some of the methods for doing so. The emphasis is on the transmission of digital data and the control of errors in a noisy communication channel. I've explained several methods of encoding the data that make modern wireless communication possible. The chapter lists and explains many of the standard tools used by communication engineers, including coding, multiuser access, security, and compression. Lastly, I've described a few of the most popular communication protocols that can be used in a robot project.

Chapter 10 covers motors. Engineers classify motors by the type of power they consume. AC and DC motors (including stepping motors) are discussed along with the different internal structures that make them work. The advantages and disadvantages of each type are presented as well.

Chapter 11 covers mechanics and covers the selection and the relevant properties of materials. Many robots have mechanical problems, so several design tips are included. In addition, short sections are dedicated to static and dynamic calculations.

PROJECT MANAGEMENT

Act 1 Scene 1:

The graying professor stands in his graying tweed suit in an overly heated classroom with high windows and ceramic tiles on the wall. The asbestos-covered steam pipes clank and bang as he stares out from behind his ridiculously high podium over a classroom of eager, young robot builders seated in hard, creaking wooden chairs. There is a long silence until his lowers his glasses, leans forward, and slowly intones the following in his best Stentorian English.

"So you want to build a robot, do you? Well, I am reminded of a wonderful scene in the movie *Young Frankenstein* by Mel Brooks. The son of the famous Dr. Frankenstein is addressed in a conversation by his proper last name pronounced 'Frankenstein.' What follows is an embarrassing, slow, pregnant pause in the conversation. The young doctor leans forward and slowly corrects his friend, 'That's pronounced "frahnkensteen."'

Just what is the fascination with robots anyway? If you remember nothing else in this book, remember this frahnkensteen phrase. Like no other, this technical field engenders passion.

It's important that you let that phrase sink in a couple of days before picking up a screwdriver. For from passion springs forces that we cannot understand. Love, joy, creativity, heartbreak, grief, and ruin all lurk to snare us as we move forward in this endeavor. And passion makes it all possible! Personally, I feel it's just as important to understand why I'm doing these things as it is to actually do them. I am old enough to realize that I will never fully understand my motives, nor should I. If I really found out exactly why I liked this field, the fantasy would probably be gone and I'd have to move on to something else.

Something is deliciously evil about trying to construct robots to carry out our bidding when we do not even know our own wishes and desires. Think about that. Have you ever seen the movie *Forbidden Planet*? It's a great, old science fiction movie partially based on Shakespeare's play *The Tempest*. I won't give away the movie's plot, but suffice it to say that a bright human gains control of a robot built by an advanced civilization. What ensues, as the robot carries out his new master's "will," is mayhem.

My point is this. Let me persuade you to stop and think first. Spend the time to analyze your motives and desires. Take the time to plan. This is not just a spiritual or psychological exercise, but it has a practical application and tangible rewards covering the spectrum from personal growth to the success of the project.

Taking this a step further, let me teach you something about the "nontechnical" art of project management first. It's a little known fact, but practicing a bit of project management makes it far less likely that your robot will run amuck and blow up the planet or that your family members will have to change their names to show their faces in public.

Project Management

Classically, a project is an endeavor to carry out some specific purpose. One English dictionary defines it as "a planned undertaking." We should note, for the record, that the Ape-English dictionary at www.ac.wwu.edu/~stephan/Tarzan/tarzan.dict.html has no entries for the words project, plan, or management. So if we are to maintain our species' lead over the apes, let's elevate our project management practices.

Why does a project require management? *Webster's* dictionary says a project requires planning. Webster did, after all, successfully finish his dictionary. Then again, we know of few people who have heeded Webster's advice in life. So let's look deeper than Webster's definition. Generally, a project has three elements: a deadline, a required outcome, and a budget. Maybe the project has no deadline, and maybe we don't know what the outcome is to be yet, but the project probably has a budget; any project always has some kind of financial limit, beyond which it will be cancelled. I'd like to make a case for having all three elements in the project.

The following discussion is based on project management processes used within a large company. The robot hobbyist, despite that fact that he or she wears all the hats in the project, should still perform the basic tasks of a *project manager* (PM). This is due to two reasons. First, the project will suffer if steps are skipped. Second, learning the art of being a PM is well worth it and will further any career.

The classic reason for managing a project is that some of the requirements will not otherwise be met. The truth is, even the most professional PMs have difficulty meeting all their goals at the same time. Half the time, a project will be late, be over budget, or fail to deliver the required results. If these goals were easy to attain, PMs would not be required in the first place. By implication, if no projects had PMs, the results would be much worse.

Many projects do not have formal PMs. Often, an engineer on the project handles some of the PM duties as a side task. Sometimes the PM duties are distributed among a few people, often with poor results. One person should be the PM and should be in complete charge of the project. That person should have all the powers and responsibilities of a PM. If you are the PM in your spare time, that's fine, as long as you can perform the tasks in the time you have to devote to the job.

First and foremost, a PM in a robot project is responsible for getting the robot done within all the restraints and requirements imposed at the outset. Certainly, a project can be executed and managed in almost any manner. To bring order to the situation, and to give all participants a clear picture of what's expected, it makes sense to use established methods and rules. The following discussion lays out the basics of project management processes but omits some of the details and reasoning to make it more readable.

Projects come in all shapes and sizes, and they are executed in all shapes and forms. This document provides a standard way to manage projects that is known to all responsible parties. It provides management tools that PMs can use to alter the course of a project and make corrections. This makes information easier to find, decreases the amount of negotiations involved, provides reliable channels of communication, and brings a level of comfort to all involved.

Project Process Flowchart

Figure 1-1 is a graphical representation of the various processes and procedures that will occur during the overall development cycle of the robot. The overall process is flexible, and deviations are acceptable as befits the situation. However, in general, deviations from the set process come at a sacrifice (see Figure 1-1).

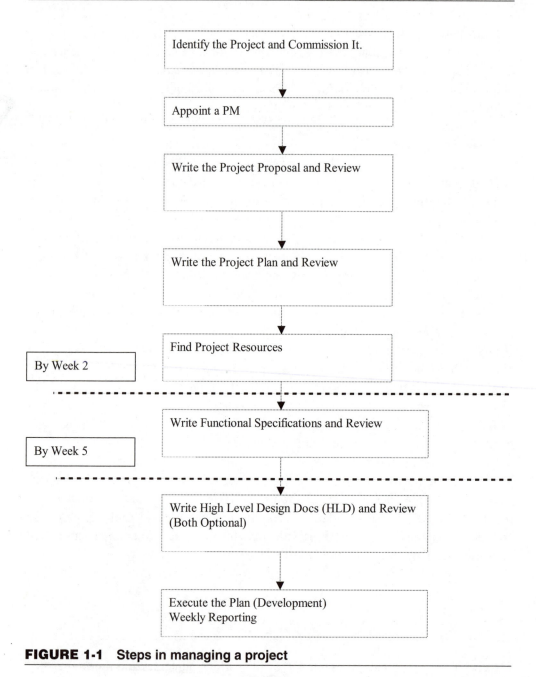

FIGURE 1-1   Steps in managing a project

How This Works When It's Implemented Right

In no particular order, these are some of the results and understandings that should come out of the proper application of this process.

The User's Manual for the "Boss"

The following words of advice pertain to the management of your robot development effort. If you are a lone robot hobbyist or operator, you are the management as well and should heed these general rules. This also applies to employees of a company involved in such a project.

PROJECTS ARE SHORT

Projects should be kept under six months. Ideally, most projects should be three to four months long at most. This means that any very long term robot projects should be broken up into a series of smaller projects. Divide the project up into functional blocks like power, chassis, control systems, and so on. This automatically engenders a complete review of all aspects of a long project at periodic intervals. By default, this includes the choice of PM, all project plans, all project resources, and so on. The following is a list of benefits that will accrue if short projects are the norm.

- PMs don't delay the project work while they get a long plan worked out. They can afford to make some mistakes over a shorter time period. These mistakes will be corrected in the next leg of the project.
- Long-term goals can be accomplished using a series of short-term goals and making corrections along the way.

PMS RUN THE PROJECTS

The PM is responsible for all aspects of the project after kickoff. "Management" might spawn the project and set the major goals, but it is the PM that runs with that information, makes a project proposal, makes a project plan (including the schedule, budget, resources, and so on.), finds the resources, executes the plan, builds the robot, and reports on a regular basis.

APPOINTING PMS

A PM must be well matched to a project. Don't overlook the fact that you might not be the right choice to be the PM! Some engineer make good PMs; others don't. The skill sets required for the two disciplines are much different.

KEEP THE PROJECT STABLE

Here are a few rules to observe:

- Don't change the tasks. Keep the specification (hereafter referred to as *spec*) stable after the project starts. The PM should give all parties the chance to change the spec up to the point when it is reviewed and development begins. If the spec must change, rewrite the project plan to accommodate the changes. Changing the spec is the second fastest way to scuttle a project's schedule and budget.
- Don't mess with the resources. Yes, this is the fastest way to scuttle a project's schedule and budget. Do not shift out resources once they have been allocated to a project. Don't borrow people, don't borrow equipment, and don't borrow space.

CORRECTIVE ACTIONS

When things are going well, about a third of all projects will still run into schedule or budget problems. Often, these projects can be identified early and corrective action can be taken. What can be done?

- Schedule a project review.
- Ask the PM for changes in the project plan, the project resources, and the project task as necessary.
- Change the PM. This is often a drastic solution, but it should not be avoided. Nor should the loss of a project be considered a significant black mark. Many new opportunities will arise for a PM to prove his or her mettle.
- Add more management. Sometimes a PM needs sub-PMs. This is often useful in large projects and can even be set up before the project starts.

The User's Manual for PMs

A checklist is provided at the end of this section that can serve as your guide throughout your robot project. The following paragraphs explain this checklist.

STARTING A PROJECT

Management currently identifies the need for a robot and determines the general requirements. This information usually comes to PMs verbally. A PM is assigned to manage the project. This assignment is for the duration of the project and is therefore temporary. In practice, a good PM is very valuable, so success in a project generally brings further appointments and further projects. However, failures will happen since the skills required to be a good PM are not the same skills possessed by even the best engineers. Being a good PM takes training, skill, and talent, and even the best PMs will trip up now and then. It is most important that you remember this. *If you are a PM and sense you are in trouble, report it to your manager.* This is the best course of action for many reasons and management should encourage it.

That said, let's assume you are the newly appointed PM of a new project. Please realize you have a large vertical *management* responsibility now. It spans sales, marketing, business, management, engineering, production, and service. Run the project like it's a business unto itself. The "business" is the best tool you have to help you succeed in your project. Use the support available for your mission: resources, space, equipment, guidance, personnel, and all other resources needed to succeed. But you have to tell management (as far in advance as possible) what is needed and what must be done. Provide all the initiative.

A further word of advice: Try to do things in the order of the following checklist. You can start portions of the project in parallel (like starting development before the plan or specification is drafted), but the risk (and potential waste) rapidly mounts. Insist on doing things in order.

RECORD KEEPING

For the moment, use the checklist to record the location of all files (documents) that are mentioned on the checklist. Keep a labeled, three-ring project notebook containing the documents and put the checklist in the first flyleaf.

PROJECT PROPOSAL

When management brings you a robot project, it generally is given in a verbal assignment. Your first task will be to write a project proposal, schedule the review meeting, circulate the proposal to the reviewers a day in advance, and preside over the review meeting. The purpose of the proposal is to crystallize thinking and estimate the costs and complexities involved. Interview the managers that commissioned the robot, senior engineers, marketing, and all other pertinent associates to obtain their opinions on all aspects of the proposal submission.

Write the proposal so someone unacquainted with the project could understand it. Describe what the robot project is, how it can be accomplished, and what resources are required. It should not take longer than eight hours of actual work (perhaps one week of elapsed time) for the interviews and for writing the proposal. The proposal is generally three to six pages long. The project proposal should have the following sections:

- **Title page** This would be something like "Project Proposal XYZ."
- **Project description** Describe to a general audience what type of robot project is needed and why it is being done. In a few paragraphs, try to describe the entire project. A simple graphic helps greatly. This section is often a page or two long, and a simple concept sketch of the robot would be appreciated.
- **Assumptions** List all the assumptions you are making that must be met for the project to go as you expect it to. This might include the existence of off-the-shelf software, timely deliveries from third parties, enabling technology, and so on. If some of these assumptions are incorrect, those reviewing your proposal can gauge your chance for success. Often, a half-dozen items are included on this list.
- **Statement of work** List all the work that will be performed during the project, with an emphasis on the largest blocks of work. The object is to acquaint the reviewers with the nature and scope of the effort required. Mention all the efforts from the initial system engineering through all the work required to *finish* the robot and document the design. Often two dozen elements make up this list.
- **Deliverables for the project** For most engineering projects, this will be the list of documents necessary to build the robot. Making this list in advance is a great way to gauge the scope of the project and to make a checklist of deliverables you can aim for. For each deliverable, estimate the delivery time when it will be finished (such as "week 7"). This will often be a list of 5 to 10 deliverables.
- **Personnel resources** This will be a spreadsheet of the people that will be needed and the total amount of labor needed from each person. The PM should pad these numbers to include the possibility that interruptions might occur. The spreadsheet should look like the following example:

PERSONNEL	WEEKS NEEDED
Hardware engineers	16
Software engineers	4
Test engineers	3
Total	23

- **Expenses** A spreadsheet should forecast any new purchases, rentals, outside expenses, and so on. It's used to budget and allocate cash flow during the project. It should look like the following example:

Expense Item	Notes	Cost
Rent oscilloscope	Three months @ $1,000	$ 3,000.00
CAD SW package	Purchase	$ 4,000.00
Outside consultant	2 MM (man month) @ $20,000	$40,000.00
Total		$47,000.00

- **Schedule** Make a table estimating the dates of major events in the project, including deliverables, major document reviews, and engineering milestones. This is just a quick estimate based on the resources and the task at hand. Another chance to estimate the schedule will occur when the project plan is submitted.
- **Acceptance test plan** How will we test the robot to be sure we are done?

PROJECT PLAN

Once the project proposal is submitted and approved, the PM should draft a project plan, schedule the review meeting, circulate the plan a day in advance to the reviewers, and preside over the review meeting.

Sometimes the project plan can be submitted and reviewed in parallel with the project proposal. The plan should be written so it can be understood and used by someone unacquainted with the project. The plan's schedule can be drawn up using a standard software package (such as Microsoft Project) in a Gantt bar chart format (about 10 to 20 bars). A portion of such a Gantt chart is shown in Figure 1-2. It's large enough to suffice for the plan at such an early stage in the project.

The plan should show milestones (with results that are demonstrable) at periodic intervals. The plan should also have a title page, an introduction, and a couple of lines explaining each task shown on the bar chart. The plan should also include a page or two explaining the approach to various issues, such as the following:

- Defusing risk, such as how and when the technical and business risks will be mitigated
- Simulation, such as how shortcuts like off-the-shelf software can be used to get moving
- Parallel execution, such as how engineers can work in parallel instead of on serial tasks
- Make versus buy decisions
- The handling of suppliers and subcontractors
- The game plan for using the personnel

The plan need not be complex and should be drafted in two hours. Two to three total pages may suffice and a partial verbal presentation is acceptable. The purpose of the review is to allow others to suggest changes in the plan that would benefit the project.

Page 1 of 2

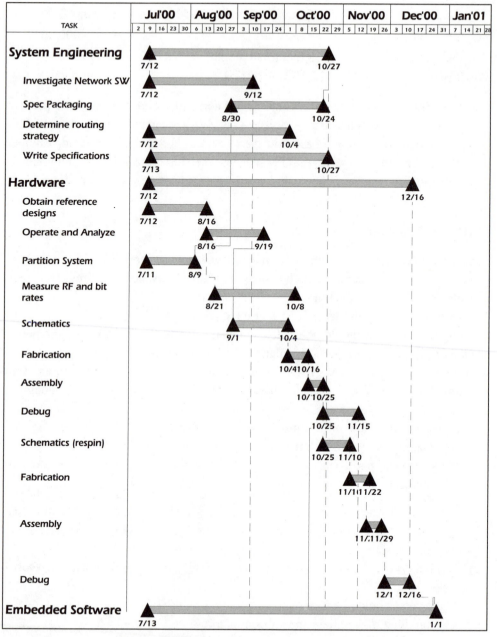

FIGURE 1-2   A Gantt chart, a standard project management tool

As the project progresses through development, it's strictly up to the PM as to how to keep track of progress and tasks, as well as the degree of tracking.

FINDING PROJECT RESOURCES

Once the project proposal and project plans are approved, work with your manager to schedule and obtain the resources you need. Due consideration should be given to possible interruptions that might occur during the course of the project. Generally, this can be done in about 10 minutes in a private meeting with management.

Human factors come into play here. Some people like to work together; some don't. PMs will draft engineers they like and can rely on. Some engineers will want to work for PMs they like and will want to avoid others they don't work well with. Some people will want to work just on the mechanics and others just on the motors or control systems of the robot. Remember, too, that people work differently. Some people can start things but will never finish. Some will never start anything themselves, but will finish things and get the project done. You will need both abilities on your team.

WRITING FUNCTIONAL SPECIFICATIONS

As an initial effort in the development project, the PM should have a functional spec written, schedule the review meeting, circulate the plan a day in advance to the reviewers, and preside over the review meeting. The functional spec is designed to fully explain the functional requirements of the robot from a high-level standpoint. The spec may take several days or longer to write and be 5 to 30 pages long.

A *system engineer* (SE), who can be appointed by the PM, typically writes the spec. The SE can be anyone (including the PM) as long as he or she is performing the SE function.

The major trick for the SE is to write requirements that best balance the needs of the reality of development cycles, the schedule, and the budget. Often, no recovery is possible if a mistake is made in this part of the project. Get the spec right first and revise them, as needed, along the way.

The spec should incorporate an outline appropriate for the hardware spec of the robot. If the robot has software as part of the design, the spec should also incorporate an outline appropriate for the software spec.

In either event, a spec document should include the following:

- **Description**   It should describe the system. To do this, just steal the proposal description. Describe to a general audience what the project is and why it is being done. In a few paragraphs, try to describe the entire robot. A simple graphic helps greatly. This section is often a page or two long.
- **Functional spec**   The spec should describe how the system should behave. It should not describe how the system must be designed or built. The design itself is up to the design engineers. All aspects of the robot's behavior should be described.
- **No repetition**   Do not repeat specifications in multiple sections of the document.
- **References**   Cite all sources and specifications that are part of the project by references. It is not necessary to repeat any part of a specification that is on file along with the spec. Use three or so words to describe the cited section and then give the section number for the cited specification.
- **Technical suggestions**   The spec can suggest specific design engineering solutions in situations where the technology is either difficult or the solution is already known.
- **Commonality**   Group common designs together. For example, if several different user interfaces will exist, consider a central body of user interface code and several different interfaces to it.
- **Unravel the toughest problems first**   The easier ones will fall into place.
- **Identify the technical risk points and elaborate on them**   This is very important since the risk items generally have the greatest chance of derailing the project. A few words of advice: *Get rid of the risk items early in the project.* In any robot project, a few risk items can bust your project wide open. They might involve the delivery of a prime component, they might involve untested technology, or they might involve personnel problems. Whatever is the case, make a plan to handle the risky aspects of the project first and foremost. Once they are out of the way, you can proceed with much more assurance and predictability.

WRITING HIGH-LEVEL DESIGN (HLD) DOCS

The design engineers have the responsibility of writing a *high-level design* (HLD) doc for the robot. However, it can be skipped at the discretion of the PM. The HLD is typically between 10 and 20 pages. The PM can schedule the HLD review, distribute the HLD to reviewers a day ahead of time, and preside over the HLD review meeting. The HLD is a technical plan showing the way the spec requirements will be implemented in the actual design. It should serve as the blueprint for the successful implementation of the final design and the HLD should make it clear how the design will be accomplished.

The following might be included in the HLD for an embedded electronics system:

- Hardware considerations:
  - Block diagrams of boards, major chips, and buses
  - Documentation (PDF files) of major chipsets
  - Power and cooling plans
  - Connectors and all package breakouts
  - Preliminary layout and plans for the board fabrication.
  - Compliance issues
- Software considerations:
  - Block diagram of major software modules
  - Performance estimates
  - Major algorithms
  - Interfaces to third-party software
  - Stack issues
  - Network issues
  - Operating system issues
  - License issues
- General issues:
  - Reference specifications (files or URLs)
  - Application notes
  - Memory map
  - Interrupt map

EXECUTING THE PLAN

Executing the plan and actually developing the robot are up to the PM and the engineers, and these tasks take up the bulk of the time during the project. Now we're up to the point where we've got a mandate to execute the project and a reviewed spec. We've got people on board and a green light to proceed. So now what? Here's some words of advice on various topics:

- **Spec**   Get all parties to read the specification and the HLD. Listen to the senior engineers (if there are any) about how to proceed. Don't be afraid to move a couple of squares backward at this stage. If any senior engineer has significant questions about the spec or any part of the project as laid out, heed them well. The best chance to make corrections occurs early in projects.
- **Leadership**   Even if you're the only person on the project, you need to consider how you will lead the project as a PM. Leadership is especially important when more people are involved. Many books have been written on the subject that you might consult, ranging from classics like Sun Tzu's classic book *The Art of War*

and Machiavelli's *The Prince*, to modern treatments like Herb Cohen's *You Can Negotiate Anything* and Scott Adams's *Dilbert: Random Acts of Management*. Things do change some, although I've run into many different leadership styles during my career. Just realize that I'm about to try to compress centuries of learning and wisdom into a page or two of usable advice.

A PM should lay out, for all concerned, the following major elements needed to lead a project:

- Vision
- Mission
- Strategy
- Tactics

The vision is a dream, a view of how the project will go and what life will be like when the project is complete. It should serve to create an image in the mind's eye for each member of the project. The image must motivate them to act with a common purpose and follow your lead.

The mission outlines the specific goals that your group will achieve during the project. Most of these goals will be directly related to the specifications and project plans. But it would be a mistake to limit your team to such goals when learning, accomplishment, teamwork, and glory are to be attained. Plan for those, too. As the old sales maxim goes, "sell the sizzle, not the steak."

The strategies are the methods of positioning and approach by which the group can achieve the individual goals in the mission. The group does *not* have to successfully accomplish the strategies, just the goals. But if the group sticks to the strategies, the goals are likely to be accomplished. The PM, with the help of the rest of the group, can determine things such as

- How hard everyone will have to work
- How to work at the same time on tasks that might otherwise need to be done one after the other
- When it's worth taking specific risks
- Which goals are more important than others

Tactics are the smaller maneuvers used to accomplish the goals of the strategies. They are somewhat different than strategies in that they can be more easily abandoned in favor of a different tactic. Several different tactics can be used while following the same strategy. The PM and the rest of the group can collectively set tactics such as

- Who works on what
- What order things are done in
- What the backup plans are if certain things don't pan out

The basic idea of the "vision, mission, strategy, tactic" thing boils down to this: Tell your people what they can accomplish, fire them up, give them a strategy so they can act together, and point them in a good direction so they can march off as an effective force to accomplish the goals.

Consider the famous speech Shakespeare wrote in *Henry V* about the English king's bloody conquest of France (the play can be found at www.theplays. org/henryV/). Kenneth Branagh played Henry V in the movie version (http:// us.imdb.com/Title?0097499) and he delivered a stirring rendition of the speech, firing up his outnumbered troops on the eve of battle as they grumbled and wished for reinforcements.

> The fewer men, the greater share of honour . . .
> I pray thee, wish not one man more . . .
> Rather proclaim . . . that he which hath no stomach to this fight,
> Let him depart . . .
> We would not die in that man's company
> That fears his fellowship to die with us.
> This day is called the feast of Crispian.
> He that shall live this day, and see old age,
> Will yearly . . . strip his sleeve and show his scars.
> And say "These wounds I had on Crispin's day" . . .
> This story shall the good man teach his son . . .
> And gentlemen in England now a-bed
> Shall think themselves accursed they were not here,
> And hold their manhoods cheap whiles any speaks
> That fought with us upon Saint Crispin's day . . .
> You know your places: God be with you all!
> *Henry V*, Act IV, Scene iii (excerpted and edited)

I actually gave this speech once to a project team, although I admit I had to read it. It was a gamble, but it was well received and had the desired effect. Notice that Henry did not try to motivate his troops by saying they could win a battle. Instead, he told them they had a great opportunity to gain glory and could tell their kids all about it. He appealed to emotions like pride, love, and a sense of accomplishment.

A leadership speech should be given at the beginning of the project to the whole team. It can be reiterated with individual team members when they need it. Further, don't forget that different things motivate different people, and individuals may need different leadership guidance.

One last thing: Let the group name the robot. The engineers will have much more personal stake in the project if they can name the robot themselves. It's almost a promise to give birth to a being of sorts.

PPP REPORT (WEEKLY)

For the project, the PM should submit a *Progress, Problems, and Plans* (PPP) weekly report for several reasons. First, it's a good record of your project. Second, you should at least have advisors who can look over the report and make suggestions. Last of all, it will light a fire under you since it's embarrassing to file an empty report even if you're the only person who reads it.

You should observe a basic rule: *Confess early and confess often.* It's much better to air problems early than to keep them secret and let them fester. They almost never get better left alone. Good management should reward PMs who turn themselves in because they have significant problems inside their project. That way adjustments and corrections can be made in a timely manner. Problems seldom get better when aged.

The weekly PPP report you write should have the format shown here:
PPP Report

Project Report: 8/15/02—Project XYZ
Progress (The most important things that happened during the week)

- _____
- _____

Problems (The most important problems that exist)

- _____
- _____

Plans (The main short-term plans to be executed soon)

- _____
- _____

PROJECT REVIEWS (SCHEDULED)

The PM should schedule regular project reviews with senior advisors and colleagues. Some reviews are called for in the project schedule and checklist. Other reviews can be scheduled for corrective action, discovery, and so on.

PM'S PROJECT CHECKLIST

A PM should fill out the following checklist during the project. This is the best way to keep track of the requirements that must be met during the project. The manager to which you report will also be following this checklist. It's your best tool to ensure that your needs will be met so you can execute the project cleanly.

Robot's name: _____

Manager: _____

Start date: _____

Targeted deliverable: _____

	Dates:
Project proposal approval:	_____
Project plan approval:	_____
Project resources assigned:	_____
Spec reviewed:	_____
HLD reviewed:	_____
PPP reports (enumerate dates):	_____ _____ _____ _____ _____ _____
Design reviews (as scheduled):	_____ _____ _____ _____ _____ _____
Acceptance test completion:	_____
Project completion:	_____

Conclusion

In summary, don't overlook the fact that a project to build a robot must be properly managed like any other. Project management is an art and a field of study in its own right.

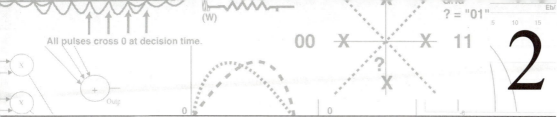
CONTROL SYSTEMS

This chapter covers the most complex topics in the book. Control systems can be very ornate and difficult to build. They can be built using computers, linear electronics, mechanical parts, biological parts, or just spit and sticks! But underlying all control systems is the queen of the sciences—mathematics. Given an understanding of the math, we can tame any of these types of control systems. In the final analysis, they all behave the same way, following the same math.

It would be heresy to some to suggest that control systems can be tamed with an understanding of just a few equations, but the fact is, the basic mathematical concepts of control systems can be greatly simplified and made accessible. If you learn the basics, you can probably extrapolate to other cases using your instinct. That's our goal in this chapter.

Control systems are everywhere and they come in all shapes and sizes:

- The average car has 35 computers in it now, running the engine, the brakes, the radio, the radar, and so on.
- You are a control system of sorts. I can surely rely upon you to turn the page when my words run off this page to the next. You are very predictable that way and you follow the western standard of page turning, as does every school kid in the country.

- Every toilet has a control mechanism for refilling the tank with the appropriate amount of water, and reliability is paramount.
- The average toaster is great at browning bread in a repeatable manner.
- You can probably walk through a completely dark room, touch a few well-known milestones, reach out with your hand, and find the light switch almost every time.

We all take the existence of such control systems for granted. Let's assume we've already built a large, strong robot body with the power, agility, strength, speed, and dexterity we believe it needs. Now comes the hard part. Here's a dream list of intangibles that might be really nice to have in the robot:

- Intelligence
- Wisdom
- Compassion
- Love
- Perception
- Communication skills

That's a long list, with many critical characteristics (that a good "person" should have) left off. How many of these things should we try to cram into the robot?

Carl Sagan, the noted astronomer and author, once commented on the intellectual horsepower inherent in the control system of an interplanetary probe. He said the probe's computer was roughly the intellectual equal of a cricket. To tell the truth, I think he sold crickets short (see Figure 2-1).

So here's a word of caution. If you hope to build a machine with wisdom and compassion, you have a huge, impossible task before you. Here are some of the profound problems you'll have to wrestle with. Forgive me for not explaining myself with all of these statements. I'd encourage you to consider each for yourself and delve into the reasons for these problems and their implications.

FIGURE 2-1   Crickets are "smarter" than many computers.

- The truth is, the human brain is capable of massive calculations, far more than the average huge computer. If you doubt this, consider the game of chess, in which humans have been beating computers for years. Computers designed for chess are only now catching up. But remember, chess is a game that a computer can at least digest easily, so the designers can optimize the computations. Most of life is much more complex than chess.

- At the risk of throwing cold water on the dreams of creative young scientists, most acts of human interaction will probably never even be defined, much less equaled by machine. Wisdom, love, and compassion spring to mind.

- The human mind has profound defects, defects that are manifest in the daily news broadcast. One could argue from an evolutionary standpoint that human defects such as those engendering greed and war are inevitable. Further, it could be argued these defects still benefit the human species and help to propagate it. It might be controversial to say so, but if we were to breed such traits out of humans, the insects would probably supplant us sooner than we might expect. As a side exercise, I ask you this. If you could press a button and make aggression, greed, envy, and other such vices instantly disappear from the human race, would you really press the button? If you could choose such traits for your robot, would you build them in?

- Humans cannot know their own minds, much less duplicate them perfectly. It won't stop us from trying though.

  - As a counterargument to my previous assertion, it must be stated that humans are having an increasingly difficult time distinguishing between human and computer "personalities." Alan M. Turing, the British mathematician famous for his code-breaking work in World War II, proposed a simple experiment that has turned into a periodic contest. The experiment, known as the Turing Test, challenges a human interrogator to hold a conversation with two unseen entities, one a computer and one a human. The interrogator must discover which is which. Winners are awarded the Loebner Prize. Visit the Loebner Prize web site for some interesting discussion and surprising results (www.loebner.net/Prizef/loebner-prize.html). More on Turing can be found at http://cogsci.ucsd.edu/~asaygin/tt/ttest.html#intro.

  - As another example of problems that cannot, and perhaps should not, be solved, consider whether your robot should be male, female, or genderless. We leave this exercise to the student body and recommend the debate be taken *outside* the classroom. A variant of the Turing Test, by the way, asks the interrogator to differentiate between a man and a woman. What questions would you ask?

- Humans cannot communicate with each other perfectly. A person can only attempt to utter the right words that will instill the proper notion of his or her idea into another person's mind. To communicate verbally, we form our thoughts, utter them, watch the reaction in the other person, and alter our statements based on his or her reaction. All these actions cannot be perfectly executed and always have unintended results. On this point, read Ronald David Laing's book *The Politics of Experience*.

A city with a large convention center suffered a flood inundating the first floor of the center. When firemen showed up at the convention center during the flood, they were amazed to hear rushing water every 45 seconds. Water gushed down the escalator from the second floor, stopped, and then repeated over and over. It turns out somebody had designed a smart feature into the elevator system. Since there were only two floors, why even bother putting in floor buttons? Just sense the motion of people coming into the elevator, and take them to the other floor. So the elevator was patiently going to the ground floor, opening up to allow the floodwater to come in, and bringing it to the second floor. Sensing a great deal of traffic, the elevator returned to the ground floor for more "people." All the while, the control system was perfectly content with its actions.

So my advice about the control system is this. Keep it simple, unless you're just experimenting and fully prepared to fail.

Let's take a step back and look at your original goals. If you've written specifications for the robot (and kept them simple), you have a limited list of tasks that the robot must perform. All you have to do is build a robot that can execute the tasks on its plate.

Where do we start designing a robot so it can do such things? For starters, we can look to nature for analogous designs. Nature abounds with control systems worthy of emulation. However, our thoughts are commonly rife with anthropomorphic visions of robots. The first image that springs to mind is of a robot with a head, two eyes, two ears, a mouth, two arms, and a torso. Are we being led astray by our own instincts?

Distributed Control Systems

Although many arguments have been made for the existence of a distributed intelligence within the human body, clearly a central control system exists: the brain. Is a central control system what we really want? This is worth considering before choosing an architecture.

Consider a school of herring. They swim in giant schools, flashing silvery in the deep blue ocean light. See http://www.actwin.com/fish/marine-pics/anchovie.mpg. As some tuna come in to attack, the school instantly swerves, divides, and coalesces as if by magic. It's a viable survival tactic for the herring. How do they pull off such a feat? Well, each individual herring simply watches his four immediate neighbors and reacts to their position, speed, and movement. The net result, observed at the school level, is dramatic and effective. Thousands of tiny brains act almost as one, and the tuna are partially frustrated. With luck, they go off to bother the shrimp.

The herring school is using a "distributed" control system. The school is governed by the collective will and common actions of the individual fish. Consider some of the advantages of a distributed control system:

- **Cheapness**   Individual control systems elements are simpler and cheaper. In this example, we'd only have to design something simple like a herring and then replicate it thousands of times (gaining economies of scale).
- **Reliable**   If the system is designed to survive the failure of portions of the system, a few failures will not bring it down. Surely, not all the herring escape the tuna. The school simply changes shape to heal up the hole where the eaten herring once was, and life goes on.

A distributed control system does have some disadvantages:

- **Communication**   Sometimes it's hard to communicate everything between individual control elements. A herring at the far side of the school doesn't know a tuna is coming until his neighbor signals such. The panic signal spreads through the school like a wave, but it might be too late. This form of knowledge truly is power, and a matter of life and death.
- **Horsepower**   The individual elements within a distributed control system generally are not powerful in and of themselves. Although the collective herring school solves the tuna problem as well as any human or computer might, the individual herring could not match a human at math or reasoning. Distributed control systems are often designed to solve specific problems and are not as good at fielding general-purpose problems. If you use a distributed control system, be very careful that you know all the problems that it must face. If the specifications change, your design might flounder!

If you'd like to explore a distributed model for robot control, here are some URLs with source software and links. Just beware; you could easily spend weeks playing with these models:

- http://www.red3d.com/cwr/boids/

The following URLs consider general-purpose distributed control systems:

- www-db.stanford.edu/~burback/dadl/
- www-db.stanford.edu/~burback/dadl/node87.html

One of the purposes of this book is to point out fields of endeavor that might lead you to a life-long career choice. If, for some odd reason, you're hooked on herring, go to Iceland (http://siglo.is/herring/en/silver.shtml)!

Central Control Systems

Let's take a look at centralized control systems. Certainly, an understanding of a single control system is vital for an understanding of a distributed control system. I'm going to leave it as an exercise to extrapolate these teachings to any work done on a distributed control system.

Most control systems are built around the same basic control structures. We'll look at a few different structures, but the point is their behavior can be described by the same math. We can discover for ourselves the sorts of characteristics that these control systems have by observing a readily available control system. The control system I've chosen to demonstrate is, right now, at the tip of your finger. We are shortly going to do some experiments while you are reading.

Open-Loop Control

Most robot control systems have some sort of input signal and output signal. In between, the control system responds to the input signal and changes the output signal accordingly. The following is a simple diagram showing an open-loop control system (see Figure 2-2).

The input signal is generally a low-level control signal. Two examples of an input signal might be the signal from the power button on a TV remote or the linear voltage from a rotating dimmer switch. Generally, in a control system, the actuator amplifies and transforms the input signal. When a person presses the power button on the TV remote, the remote generates an infrared signal that the TV interprets to close a relay and give

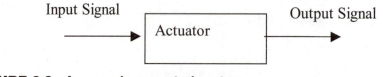

FIGURE 2-2 **An open-loop control system**

power to the TV circuits. Actually, two open-loop control systems are at work. They are concatenated and operate as a single open-loop control system (see Figure 2-3).

In open-loop control systems, the information tends to flow only one way. For example, the control system inside the remote never finds out if the TV goes on or not. Furthermore, the power button on the remote never indicates if the infrared beam was sent out or not. If your finger is over the optical opening, nothing happens at all and the remote never knows the TV has not gone on.

Let's run an experiment illustrating an open-loop control system within your body. Glance over to your right and locate an object in the room. Remember where it is and then look back here to the book. Now close your eyes, point to the object, trying to put your finger right on the object in your field of vision. Open your eyes, and see how close you came (see Figure 2-4).

You'll notice that you never really get it right with your eyes closed. When you open your eyes, you can see your finger is a little off. The error will never go away and is called the steady state error. It's an error that will persist long after the control system has settled on the final output and will make no further corrections. We'll see steady state error as a term in the equations that we develop later. All control systems have this error. It's an important parameter because when you are designing a control system, you must keep the steady state error below acceptable bounds.

You can perform another experiment if you have a dimmer in your home. Wait until dark and turn off the dimmer, making the room dark. Close your eyes and then turn on the dimmer to where you think the minimum acceptable reading light level is.

Power Butt   **Infrared Signal**   **Power to TV**

```
          ┌──────────────┐        ┌──────────────┐
  ───────▶│  TV Remote   │──────▶ │     TV       │ ──────▶
          └──────────────┘        └──────────────┘
```

FIGURE 2-3   Concatenated open-loop control systems

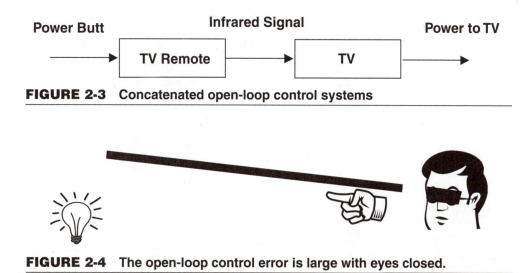

FIGURE 2-4   The open-loop control error is large with eyes closed.

Open your eyes and see how well you did. Likely as not, you won't be satisfied with the light level because the steady state error will be too large. You will have to make a correction in the light intensity to be comfortable reading under the light.

The corrections you have made in these two experiments by finally using your eyes illustrates an important concept. An open-loop control system can be improved if it is told how well its output matches the input requirements. With that somewhat broad statement, we'll introduce another type of control system.

Closed-Loop Control

Closed-loop control systems are also referred to as feedback control systems, because information flows backwards at some point within the control system. Generally, this reverse information flows from the output of the control system backward toward the input. The information that flows backwards allows the control system to make corrections in its output. Figure 2-5 is a generalized diagram of a simple closed-loop control system.

Information flows backwards in the system, from the output signal to somewhere near the input. I've labelled this reverse information flow "feedback." In this simple version of a close-loop control system, the output signal is sent back and directly compared to the requirements set by the input signal. The circle shows an arithmetic computation (subtraction). If the output does not directly match the input, the actuator will receive a nonzero signal at its input and provide corrections at the output so its input returns to zero. In practice, many different kinds of closed-loop control systems exist, and, as such, one could make many variations to this diagram.

Many control systems do not have outputs that are directly comparable to the inputs; the circle in Figure 2-5 must be much more complex than a simple subtraction element. Often, the output signal must be transformed before it can be compared to the input sig-

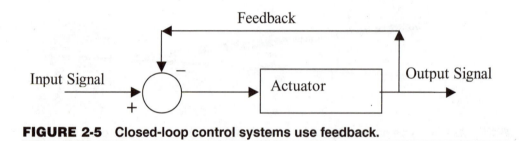

FIGURE 2-5 Closed-loop control systems use feedback.

nal. Such transformations may take the form of scaling (to a different size) or conversion from one signal type to another (like from light values to a voltage signal).

Often, the comparison within the circular symbol is not a simple subtraction. Sometimes it's a comparison (bigger or smaller) and the output of the circle represents either off or on. Thermostats work this way, for example.

Clearly, the system looks like a closed loop. Often, such a system is also referred to as a closed-loop feedback system. All these terms generally mean the same thing.

Let's run the first experiment over again a different way as a closed-loop control system. Now close your eyes and point again to the object (trying to put your finger right on the object in your field of vision). Open your eyes again, and see how close you came. You still didn't get it right with your eyes closed, but now with your eyes open, you've introduced feedback into the system. With your eyes open, it's easy for you to make the correction and get your finger right over the object in your field of vision (see Figure 2-6).

Notice, the steady state error is now much less. We think the error is actually zero, but we'll see shortly that this is rarely the case. Certainly, closed loop control is a better solution in terms of accuracy, but it comes at the cost of providing extra control elements (in this case, vision).

STEADY STATE ERROR

Now that we've identified a parameter of interest, let's look at the math. We can assign arbitrary variables to represent the signals and control system elements that we have been talking about (see Figure 2-7).

Looking at the circular arithmetic element (subtraction),

$$b = a - d$$

The actuator is said to have a gain of C. This gain can be immense and the system will still work. As an example, if a very tiny positive signal takes place at b, then signal d can be extremely large and positive. Similarly, if a very tiny negative signal is issued at

FIGURE 2-6 The closed-loop control error is smaller with eyes open.

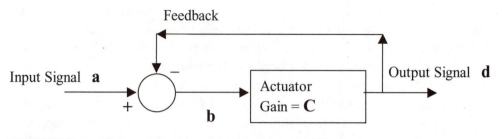

FIGURE 2-7 A closed-loop system with an actuator and error signal b

b, then signal d can be extremely large and negative. The system is designed to function with signal b being very small, nearly zero. The actuator usually provides the horsepower and amplification to drive signal d. To be precise,

$$d = C \times b$$

Substituting for b in the previous equation, we get

$$d = C \times (a - d)$$
$$d = C \times a - C \times d$$
$$d + C \times d = C \times a$$
$$d \times (1 + C) = C \times a$$

Finally, we have the relationship between the input a and the output d:

$$d = a \times (C/1 + C)$$

This equation predicts that the steady state error of this sort of closed-loop control system is governed by C. The output d will be off by the ratio of $C/(1 + C)$. This factor is also termed the steady state error coefficient. Note that it cannot be zero; a steady state error always exists. Note also that the larger the gain, C, of the actuator, the smaller the steady state error. As C tends toward infinity, the steady state error tends toward zero. What practical things can we do with this math?

- Expect the closed-loop control system to exhibit some steady state error. Don't be surprised if the system does not exhibit a perfect output. It is bound to have some error.
- Recognize that the steady state error is very likely to depend upon the gain of the actuator. Use the steady state error coefficient to estimate what that error will be in advance and design the robot to allow for an error of that size. If the system has too much steady state error, consider revising the actuator gain to correct it.

■ We might be led to believe that making the actuator gain as large as possible is desireable. Just be aware that increasing the gain of the actuator adds expense and will adversely affect the dynamic (nonsteady state) behavior of the control system as we will see later. In the worst case, a large actuator gain can make the system unstable and lead to failures. Whenever altering the gain, remember to reevaluate and retest the dynamic performance of the control system.

Realize that these equations model a general-purpose closed-loop control system. If the control system is meant to control the robot's position, then the variables a, b, and d are measured in distance. If the control system is meant to control the robot's speed, the variables are measured in speed. If the control system is meant to control the robot's acceleration, the variables are measured in acceleration. The fundamentals of the math are still the same; only the units change. We can use the equations herein to control any of the aforementioned systems without further investigation.

We leave it up to the reader to investigate the mathematics of calculus that hold that acceleration is the derivative of velocity, and velocity is the derivative of position. Suffice it to say that positive acceleration builds up speed, negative acceleration (braking or accelerating in reverse) decreases speed, positive speed accumulates distance (position), and negative speed (moving backwards) decreases distance (position).

DYNAMIC RESPONSE

When a control system sees a changing input, it generally changes the output. A standard test of a control system is to give it what's called a step input. For a robot, such an input might call for it to move from its present postion to a new position and stop there. The classic input used to test a control system is a step input and is of the following form (see Figure 2-8).

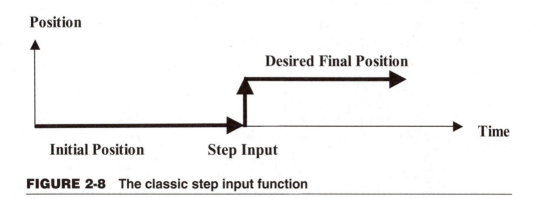

FIGURE 2-8 The classic step input function

An ideal control system would follow the step input function and produce the same step output function. The robot would instantly move to the new position and stop on a dime with no steady state error. We've already seen how the robot will have a steady state error (not fully reaching the desired final position). The truth is, the robot cannot move instantly and it cannot stop on a dime. The control system in the robot sees the step input, delays a bit for reaction time, finally starts moving, and tries to stop near the final position. The response is going to be imperfect no matter which way we slice the pie.

So before we look at how control systems really behave, we're going to have to stop and do some math. After that, we'll have to tools to see the following:

- How the design of the control system determines how the robot will react
- How to characterize the robot's performance in a few parameters
- How to know which design parameters to alter based on the robot's performance
- How to get optimum performance from the robot

To get the tools we need to analyze and manipulate the performance of the robot, we're going to pick a mathematical model for the robot and derive some of the equations. We're going to skip the easier models of robot behavior and go straight to a slightly more complex case. We are going to use math and physics that might be beyond the casual reader's abilities, but we will return to a usable, intuitive model of what's going on. We will start with physics, calculus, Laplace transforms, and algebra to arrive at usable results. Once we have that math in front of us, we will explore the tools it affords us.

First, we need a way to look at the parts of the robot and assign numbers to the movements we observe. This can be done in a couple of ways:

- **Energy evaluation**   One way to analyze dynamic movement is by looking at everything in terms of energy: where it's stored and how it's used. We are not going to use this technique any further in this book, but it's worth mentioning the alternate technique. Energy is stored in multiple places in a robot, certainly in the batteries, but it is also temporarily stored in other places
  - **Springs (potential energy)**   A good mathematical description of springs will be provided a little further along in this book. As a spring is compressed, the energy E in the spring is

$$E \ = \ 0.5 \ \times \ K \ \times \ x^2$$

where x is the compression distance. Note this equation only works for smaller values of x because an overly compressed spring becomes nonlinear and runs out of springiness. K is the spring constant; a bigger, stronger spring has a larger K value.

■ **Moving mass (kinetic energy)**   The energy in a moving mass is

$$E = 0.5 \times m \times v^2$$

where m is the mass, which will be described later in the book. v is the velocity. Note that a moving mass might not just be moving linearly. It might also be rotating. As such, you can model the energy of both motions separately. You can use the center of gravity of the mass and see how fast that is moving linearly. Then you can add the energy of rotation about that center of mass (as you find it).

■ **Mass at heights (potential energy)**   When a mass is at a height, the potential energy it has is given by the equation

$$E = m \times g \times h$$

where M is its mass. g is the acceleration constant of gravity (32 ft/sec^2). h is the height the mass might fall. A nice treatment of potential and kinetic energy can be found at www.dcate.net/coasters/pe.html.

■ **Force evaluation**   Instead of looking at energy, we're going to use the technique of looking at everything in terms of force. We need only to characterize the forces within the system as they act together. In this way, we can predict what the pieces of the robot will do. Here are some of the places force is stored in a robot:

■ **Motor force**   Most motors will generate a time-varying force when energy is applied. The force might be rotational or linear. To keep matters simple, we'll be looking at linear force, such as might be applied by a solenoid, which is an electromagnet with a moving metal core, much like Figure 2-23.

■ **Moving mass force (kinetic force)**   Newton created the equation for force acting on a mass (or mass creating a force):

$$F = m \times A$$

where m is the mass and A is the acceleration (or deceleration). When gravity is the force providing the acceleration, A = g and thus F = m $\times$ g, the force needed to hold up a mass m.

■ **Spring force**   A spring with a spring constant of K will have a force

$$F = K \times x$$

where x is the compression (or elongation) of the spring.

■ **Friction force**   Friction is a force that is engendered by velocity through a friction medium. For example, a motor, when the power is turned off, will cause the vehicle to coast to a stop because its rotor glides over the bearings and the grease in the bearings still has friction. The decrease in speed is somewhat linear in time. Friction is proportional to velocity and has a force of

$$F = B \times v$$

where B is the coefficient of friction, and v is the velocity. This makes intuitive sense. When you rub your hands together, you have to work harder to rub faster. The friction grows hotter the faster you go. The force increases and the energy mounts up faster.

Friction comes in disguised forms. We often think of friction as something dragging over a surface. Often, elements will have their own internal friction. A motor will coast to a stop by itself. Springs will heat up as they bounce and will slowly stop bouncing by themselves. If the coefficient of friction is not specified inside a system, you can often determine it empirically. The quick way to do so is to compute the instantaneous deceleration of a mass and compare the two forces:

F = m × a for the mass

F = B × v for the friction, so

B = m × a/v

This technique works for rotational, linear, or spring-type movements.

So now we have to pick a mechanical model of the robot in order to make a mathematical model for it. We will pick an arbitrary model that will probably be different than our robot's actual mechanics. However, once we learn how to analyze and manipulate this arbitrary model, it will be second nature for us to extend our knowledge to other models. Most systems, even unusual nonlinear ones with spasmodic motions, can be treated similarly to the model we will study. The math is close to the same. We are looking at what is called a second-order system, so called because the forces are based upon three different representations of positions (as represented in terms of calculus):

- **Position**   The position, x, of a mass. For springs, the force is proportional to x.
- **Velocity**   v, the rate of change of position, x, of a mass, the first derivative of x. In calculus, this is called the first derivative of x with respect to time (v = dx/dt). In everyday terms, we think of it as miles per hour. The force of friction is proportional to dx/dt.
- **Acceleration**   a is the rate of change of velocity, the first derivative of v, the second derivative of position x. In calculus, a = dv/dt or, when written in terms of x, $a = d^2x/dt^2$.

In a simple system where the acceleration is a constant (such as gravity acting on a falling object near the surface of the earth):

- $v = a \times t$
- $x = 0.5 \times a \times t^2$

The simplest second-order mechanical model is a weight hanging from a spring. Since almost everybody has performed this experiment as a kid, let's think back to how this system behaves. We're going to diagram the behaviors, one at a time, and enumerate the behaviors so we can explain them later once we have the equations:

1. When you displace the weight (mass) vertically and let it go, it will bounce up and down at a nice constant frequency. If the displacement keeps the spring in its linear region (without compressing it or stretching it too much), the motion of the mass will be like a sine wave. To try this, hang a weight from a rubberband until the rubberband is half stretched out. Pull the weight down a little and let it go. It will bounce up and down with a fairly fixed frequency and look like the sine wave in Figure 2-9. This illustrates the resonant frequency of the second-order system, which we will later call ω. The frequency ω is measured in radians per second where there are $2 \times \pi$ radians in a single cycle.

2. We know that if we put a bigger weight on the spring, the weight will bounce up and down slower than the lighter weight does. To try this, hang two weights from the rubberband. This illustrates how ω decreases with the mass m (see Figure 2-10).

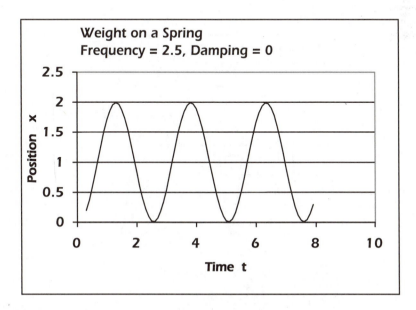

FIGURE 2-9 The movement of a weight on a spring

3. We know that a stronger spring will make the weight bounce up and down faster than the weaker spring. To try this, put a second rubberband right next to the first one so that they act in unison and use the original single weight. This illustrates how ω increases with the spring constant K (see Figure 2-11).

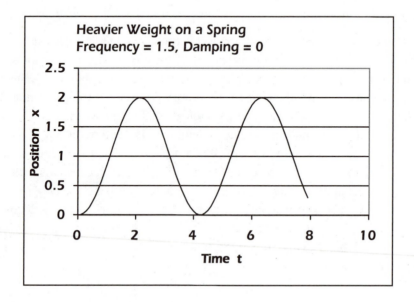

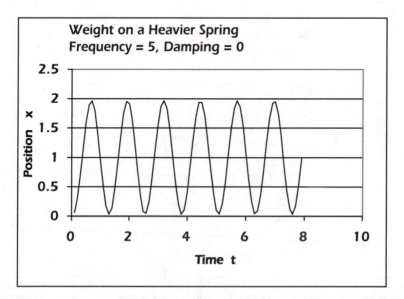

FIGURE 2-11   The movement of a weight on a heavier (more powerful) spring

4. We know that the bouncing weight will eventually settle down and stop bouncing if we stop moving the spring. This illustrates the damping action of friction. In this particular case, the friction is inside the spring itself (and in the air). The rubberband heats up as the friction inside the rubberband uses up the energy that was in the moving weight. Later we'll call the damping coefficient δ. Clearly, if you try this experiment underwater instead of in the air, the friction would be much larger and the system would settle down much faster. (see Figure 2-12).

5. We know that, as we move the top of the rubberband up (like our step input diagrammed earlier), the weight will shoot higher than the desired final position and will eventually settle down to a higher level. We call this excess movement of the weight the overshoot (see Figure 2-13).

Now it's time to diagram our model mechanical system. Instead of a hanging weight, we're going to eliminate the force of gravity and use a horizontal system where the weight rests on a slippery surface. If you want to take this horizontal system and extrapolate it to a vertical system, just extend the spring to counteract the force of gravity's acceleration on the mass. For our computations, the horizontal model takes this term out of the math since gravity does not stretch the spring (see Figure 2-14).

The ground reference is, in this case, the earth. It's not supposed to move under you (those of you in California take note). In reality, as you walk one way, the earth rotates the opposite way. But since it's so much larger than you, the motion is imperceptible. I

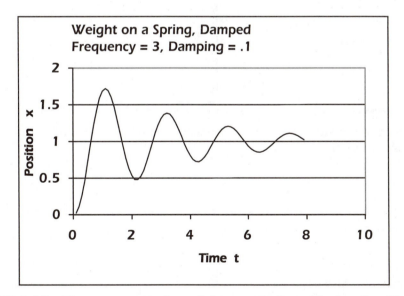

FIGURE 2-12 The movement of a weight on a spring with damping friction

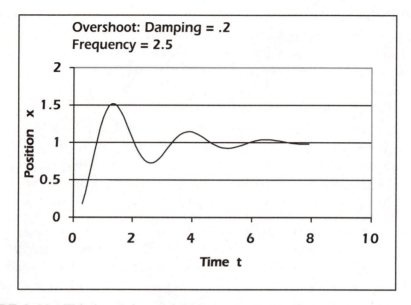

FIGURE 2-13 This bouncing weight overshoots by 50 percent

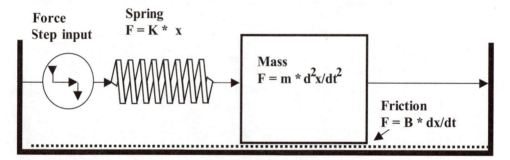

FIGURE 2-14 A second-order mechanical system with step input, spring, weight, and friction

leave it up to the reader to calculate the rotation of the earth that would occur if everybody on Earth starting walking in the same direction at once. For now, let's consider the ground stable.

We're going to delve into physics and math here without a serious attempt to explain how things work. Take heart that we will return to more familiar ground shortly and that the results will be intuitive and usable.

The force in a closed loop of mechanical elements adds up to zero. From this, we get the "characteristic" differential equation of this mechanical system:

$$m \times d^2x/dt^2 + B \times dx/dt + K \times x = 0$$

This says the spring force acts trying to accelerate the mass and overcome friction.

In calculus, many ways exist for solving a differential equation like this. The mathematics get a bit difficult, but French mathematician Laplace provided a shortcut in the form of his Laplace transforms. They basically eliminate the requirement for integral calculus and reduce the problem to algebra and searching some tables. We will perform a Laplace transform on our differential equation, do some algebra, and then use the tables to perform an invervse Laplace transform to get back our real-world answer (see Figure 2-15).

First, we transform our differential equation using the methods of Laplace. Substitute the variable s to stand for a single differentiation. As such, the differential equation becomes

$$M \times s^2 + B \times s + K = 0$$

We're going to use algebra to find the roots of this quadratic equation. Remember the old formula for finding the roots of the quadratic equation? I bet you thought you'd never use it! Stay awake in school! The following restates the quadratic equation and

FIGURE 2-15 Pierre-Simon Laplace

shows the two roots. Notice that the two roots are shown with a +- notation in the following sections:

- $a \times x^2 + b \times x + c = 0$
- $x = (-b +- (b^2 - 4 \times a \times c)^{1/2})/2 \times a$

We are going to use the quadratic equation to solve our characteristic equation. First, we are going to cheat a little, because we already know the answer. We're going to change some of the constants in the characteristic equation before solving for the roots. This allows us to easily see the final result. Here are the three changes we make:

- Divide by K so

$$m \times s^2 + B \times s + K = 0$$

changes to

$$m/K \times s^2 + B/K \times s + 1 = 0$$

- Substitute $1/\omega^2$ for m/K. Take a look at the second and third behaviors (Figures 2-10 and 2-11) of the bouncing weight we showed above, and you'll start to appreciate this substitution.
- Substitute $2 \times \delta/\omega$ for B/K. The damping coefficient δ, integral to slowing down the system over time, is directly related to the coefficient of friction, as we might expect.

The equation changes with the substitution from

$$m/K \times s^2 + B/K \times s + 1 = 0$$

to

$$(1/\omega^2) \times s^2 + (2 \times \delta/\omega) \times s + 1 = 0$$

Using the quadratic equation, the two roots are

$$s = (-(2 \times \delta/\omega) +- ((2 \times \delta/\omega)^2 - 4 \times (1/\omega^2) \times 1)^{1/2})/2 \times (1/\omega^2)$$

Take out the factors of 2:

$$s = (-(\delta/\omega) + - ((\delta/\omega)^2 - (1/\omega^2))^{1/2})/(1/\omega^2)$$

Multiplying the top and bottom by ω^2 brings us to the two roots of the quadratic:

$$s = -(\delta/\omega) + - ((\delta^2 - 1)^{1/2}$$

Now we perform the inverse Laplace transform using the tables (which are not replicated herein). For the cases where δ is less than 1, we have what's called an underdamped system that responds much like the overshoot chart. In this case, the Laplace tables show the basic solution to be of the form

$$x(t) = 1 + c1 \times (e^{(-\delta \times \omega \times t)}) \times sin(\omega \times (1 - \delta^2)^{0.5} \times t + c2)$$

where c1 and c2 are to be determined by initial conditions. To find the initial conditions, we look at the equations for the rest state of x and dx/dt. This gives us two equations in two unknowns and leads to the final solution, which is

$$x(t) = 1 - (e^{(-\delta \times \omega \times t)}) sin(\omega \times (1 - \delta^2)^{0.5} \times t + acos(\delta))/(1 - \delta^2)^{-0.5}$$

This is the final solution and was used to generate the charts earlier. This equation represents a unit step function starting from x = 0 at time 0 and settling at the value of x = 1 after the transients settle out. This can be seen in the behavior of the individual functions in the solution. The exponential function $e^{(-\delta \times \omega \times t)}$ dies off over time as t goes to infinity. The larger the damping, the faster it does so. The function

$$sin(\omega \times (1 - \delta^2)^{0.5} \times t + \ldots)$$

oscillates and provides the ringing.

Designing the Control System

Well, we've come through the math gauntlet and come up with a closed solution of how the model system behaves. Now, how do we make this useable? Remember our goals; we are going to answer the following:

- How the design of the control system determines how the robot will react
- How to characterize the robot's performance and which design parameters to alter
- How to alter the robot's design parameters
- How to get optimum performance from the robot

Let's tackle the first goal.

HOW THE DESIGN OF THE CONTROL SYSTEM DETERMINES HOW THE ROBOT WILL REACT

We have made a model of a second-order system and have the closed equation describing how the model behaves. If we know m, K, and B, we can graph the theoretical behavior of the system. Here's a step-by-step method of doing just that:

1. If you have values for m, K, and B, skip ahead to step 2.

    a. Mass   To measure the mass m, just weigh it in kilograms and divide by the gravitational acceleration of 9.8 m/sec². It should be mentioned here that kilograms is *not* a measure of weight. The actual unit of weight in the metric system is the Newton! It is *not* correct to report weight in kilograms. You should be aware that mass is not the same thing as weight. Mass is a measure of the amount of "stuff" in the object. Weight is a force and is a measure of the force exerted by the mass in the presence of the gravity created by another mass like the earth. Mass in orbit is weightless, yet retains its mass. Mass on Earth becomes weight because it's acted upon by the acceleration of gravity $(F = m \times g)$. Here's a web site about this matter:http://feenix.metronet.com/~gavin/physics/wgt_mass.html.

    This brings up an important point. The calculations for the model's second-order system are partially dependent upon gravity. The robot might not work the same way in orbit. The friction we diagrammed in the model's mechanical second-order system depends on the friction of the mass resting on a surface. Without gravity, there will be no such frictional coefficient B to speak of. You can introduce other friction elements into your robot design that would work in orbit, such as a piston with a viscous fluid within it (like a shock absorber).

    b. Spring constant   To measure the spring constant K, hang a known weight from the spring without stretching it too far. The ratio of the displacement of the spring to the weight will give you K using the formula

$$m \times g = K \times displacement$$

where $g = 9.8$ m/sec², the acceleration of gravity. The example given at the web site www.iit.edu/~smile/ph9013.html cites a 250-gram weight suspended from the spring.

$$Solving\ m \times g = K \times displacement$$
$$250\ grams \times 9.8\ m/sec^2 = K \times displacement$$
$$K = (2.4\ kgm/sec^2)/displacement$$
$$K = 2.4\ newtons/displacement$$

Hang the 250-gram weight, measure the displacement in meters, and then compute K in newtons per meter.

c. Coefficient of friction

    i. First, you must know how friction behaves, since it can get complex. The friction is greater in our model when the weight is not moving. This is termed static friction. Once the mass starts to move, the friction decreases to a lower level as long as the mass continues to move. Think of friction as a series of microscopic speed bumps. They don't seem as bumpy if the weight is moving faster, but if the weight slows to a crawl, the speed bumps are painful to go over. We've all experienced static friction before. Often, it takes an extra heave-ho to start pushing something, and a bit less effort to keep it going. Just be aware that system behavior won't precisely follow the model if B is greater when the mass is at rest. A couple of web sites about friction are located at www.iit.edu/~smile/ph9311.html and www.iit.edu/~smile/ph9104.html.

    ii. The coefficient of friction B can be measured in two ways:

Force conversion: Take a spring with a known spring constant K and use it to pull the weight at a constant velocity dx/dt across the friction surface. The force exerted by the spring is K $\times$ x, where x is the displacement of the spring. At a constant velocity, the spring force equals the force of friction, which is B $\times$ dx/dt.

$$B = K \times x/(dx/dt)$$

Derivation: We'll see later how, knowing K and m, we can derive B by observing the system behavior. This would prove useful when changes have to be made to either of the three parameters to change system behavior.

2. Let's assume we know B, K, and m. We can plug these numbers into the equation for x(t) and plot the predicted results. The robot should follow the model's behavior if the model truly does mimic the design of the robot.

Let's tackle the second goal.

HOW TO CHARACTERIZE THE ROBOT'S PERFORMANCE AND KNOW WHICH DESIGN PARAMETERS TO ALTER

Figures 2-12, 2-13, 2-16, and 2-17 were made using Excel spreadsheets. They show the predicted behavior of the model's second-order system. The figures were made specifically to show how we can guide the design and make the robot behave the way we want it to. This, of course, is the third goal, so we'll postpone that part of the discussion.

Every individual curve in the figures represents the predicted behavior of a second-order control system given specific design parameters that are affected by B, K, and m. Every curve on the figures is normalized and shows a control system that will eventually settle to the value of 1. Because of the design differences (reflected in each curve), they behave differently. The key for us is to learn how these curves behave and how to control them.

The first thing to notice about the two figures is the predictability of the curves. In Figure 2-16, marked Varying Damping Only, we can see that all the curves have about the same frequency. The center horizontal line represents the final value of 1. All the curves cross the center line at about the same times: 2.5 seconds, 4 seconds, 6 seconds. This is because each of those second-order systems was designed to have the same frequency. These curves show the effect of changing the damping.

In Figure 2-17, marked Varying Frequency Only, we can see that all the curves have about the same overshoot and undershoot. They all rise to a value of 1.5, drop to a value of 0.75, and so on. This is because each of those second-order systems was designed to have the same damping. These curves show the effect of changing the frequency.

We will examine the characteristics of the curves on the graph and discuss which characteristics are of immediate interest. Robot designers consider the following:

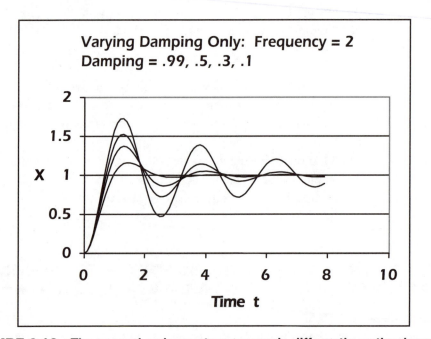

FIGURE 2-16 The second-order system responds differently as the damping is varied.

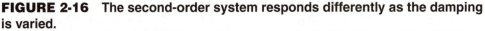

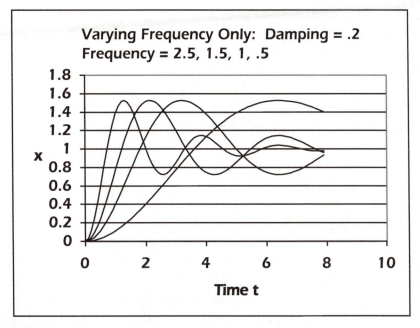

FIGURE 2-17 The second-order system responds differently as the frequency is varied.

■ **Response time** Take a look at Figure 2-17 entitled Varying Frequency Only. It was made holding the damping parameter δ constant and varying the frequency ω (we'll get into how to do that soon). The point is, the curves rise toward the final value of 1 at varying speeds. A few ways are available for measuring the response time, including

■ Time from 0 to first crossing of 1
■ Time from 0 to first peak overshoot (the maximum value)

The system has a different response time for different values of damping. If we look at the time from 0 to the first crossing, the four curves vary in rise time from 3/4 seconds to almost 4 seconds. These four curves vary in frequency from 2.5 to 0.5 radians per second. A circle has 2π radians. Frequency is related to radians in the following way:

$$F = 2 \times \pi \times \omega$$

where F is in Hertz (cycles per second), ω is in radians per second, and π is 3.14159 . . . .

Considering a cycle contains $2 \times \pi$ radians, the four curves represent frequencies of 0.4 to 0.08 Hz and periods (1/frequency) from 2.5 seconds to 12.5 seconds. Let's look at a table of some of these values and see how they relate to the response time.

Frequency, radians	2.50	1.50	1.00	0.50
Frequency, Hz	0.40	0.24	0.16	0.08
Period, seconds	2.5	4.16	6.25	12.5
Time from 0 to 1 (T0-1)	0.7	1.20	1.80	3.60
Ratio of T0-1 to period	**0.28**	**0.28**	**0.28**	**0.28**
Time from 0 to first peak	1.3	2.1	3.2	6.4
Ratio of T0-peak to period	**0.52**	**0.50**	**0.51**	**0.51**

Here are two usable rules of thumb. These numbers help you make sure the system responds fast enough to suit your requirements:

- The response time from t = 0 to the curve reaching a value of 1 is about 28 percent of the period. The period can be computed from ω as detailed just above. This allows you to pick your rise time as you pick ω.
- The response time from t = 0 to the first peak is about 51 percent of the period (as you might expect from a sine wave).

- **Overshoot** Take a look at Figure 2-16. It was made holding the frequency ω constant and varying the damping constant δ (we'll get into how to do that soon). The curves overshoot the desired level by different amounts. The smaller the damping, the larger the overshoot. Overshoot can be important because it might cause your control system to lose track of the final target. Remember the robot competition we spoke of in the introduction? The robots were all too powerful and were zipping over the control line so far that they wandered out of the sensor range and became lost. That was too much overshoot.

- **Settling time** You might think that increasing the damping is always desirable in order to decrease the "ringing" and make the system settle down faster. Take a look at Figure 2-16 to see this occurring. Certainly as the damping increases, the system looks less wild and converges to the final value of 1 faster, but look at the response time. As we increase the damping, the response time increases also, so you will have to make a tradeoff to fit your robot's design. Damping is about the only parameter you can increase that will improve the settling time.

- **Frequency of oscillation** Sometimes the control system will be even more complex than a second-order system. Sometimes the mechanics or electronics are sensitive to specific frequencies of oscillation. This can happen if the mass in the model has a resonant mechanical frequency. Remember the bridge called

Galloping Gerdie? It shook itself to pieces because the mechanical engineers missed damping out a resonant mechanical frequency. Talk about a failure to control damping. See www.ketchum.org/tacomacollapse.html for an interesting treatment of this particular mechanical failure.

■ **More variables** This brings up a good point. All along, we have assumed that both the mass and the friction beneath the mass are fixed with respect to frequency as the position of the mass changes.

If the mass is not solid but has a harmonic resonance in its structure, then the system will *not* behave per the model. So be very careful that your robot has a solid construction and as few resonant mechanical elements as possible. It is much easier to control the position of a one-pound block of steel than it would be to control a one-pound bowl of jello.

If the coefficient of friction varies with position, similar problems could occur. We have to clearly identify all the frictional elements at work within our robot system. Some will be inherent in the materials (like in the springs). Other frictional elements will be accidental and must be carefully analyzed to make sure they stay constant with position. Its not wise to allow unspecified frictional elements to govern our system. To take back control of the design, we can can deliberately put a frictional element of our choosing into the system. If it is much larger than the inherent or accidental frictional elements, it will swamp out their effect as much as possible and make our design more reliable in its performance.

■ **Stability** An entire body of control system theory is devoted to the stability of systems. We certainly know from the bridge example that it's important. It's also extremely complex in the mathematical theory and we need not go into it here, but we should look at several pieces of advice. First, we should identify just what instability is.

Some control systems, if not designed right, can oscillate way too much, upset the mechanics, and ruin the operation of the robot. These oscillations can stem from various flaws in the design.

■ **Resonant frequencies** As we just mentioned, make sure the mechanics and other physical elements of the system, such as the frictional components and spring elements, do not have resonant frequencies. Make sure they behave the same way across all the frequencies to which the robot will be subjected. One way to ensure this is to put the system on a mechanical vibrator, as we'll see later.

■ **Bad selection of the frequency** ω Sometimes the mechanical system does have some resonant frequencies within the design. If ω is chosen wrong, the ringing may be way too large and the system may be unstable. Alter ω and see if things calm down. If this helps, then analyze the mechanics again.

- **Nonlinear elements** We have to realize that our model depends on a linear behavior of all the components. We expect a smooth performance all the way around. Between loose pieces (that might move free and then snap tight) and some "digital" elements (that are on-off), some jerky motion will occur. Try to minimize the effect of these components; we'll look at nonlinear design in a while.

- **Too much overshoot** Sometimes a system will move the robot too far and be unable to recover. Such a situation occurred in the introduction where a robot moved too far in one single motion and its limited "eye" was not given time to see that it passed the boundary where it was supposed to stop. Such a situation can occur if there is too much overshoot. One solution is to increase the damping on the system.

- **Complex designs** Often, the robot is much more complex than our second-order system. If it really is a third-order or higher system, take the time to try to simplify it. Look at the performance and look at the specifications.

Let me give you an example of trouble brewing. Suppose we are trying to design a baseball robot. It has to run, catch, and throw. It might be able to run and catch at the same time, but it would be simpler to build a robot that would run under the ball, stop, and then catch it. Similarly, it would be simpler if the robot would stop running before it had to throw the ball. Granted, a human baseball player would never get to the majors playing like that. However, if the specifications and performance requirements can be relaxed ahead of time and if we can afford to have a clunky robot player, then our design will be much simple if you can partition the design. We then just separately design a runner, a catcher, and a thrower. We do not have to combine the designs and suffer the interactions that drive up complexity and threaten the stability of our design. Again, we repeat the old advice: Keep it simple.

You laugh about robots playing baseball? Just keep your eyes on the minor leagues! See Figure 2-18 from http://home.twcny.rr.com/mgraser/ballpark.htm.

So how do we stabilize a system? Several symptoms can occur. They're easy to observe and correct:

- **Severe overshoot** Sometimes overshoot can become very large. We can fix it by increasing the damping constant δ (we'll get to how that's done soon). Refer to Figure 2-17. Changing ω won't affect the overshoot much. If changing doesn't help, perhaps the robot is not following the model and we should determine why.

- **Severe ringing (the oscillations are causing problems)** To fix this, we can increase the damping constant δ. This will help decrease the oscillations sooner. If the oscillations are still objectionable, we must investigate why this is the case.

FIGURE 2-18 A baseball pitching robot trying for the Cyborg Young Award

If the robot is susceptible to oscillations at specific frequencies, consider altering ω to a frequency that might work better inside the system.

■ **Unknown oscillations** Sometimes robots will just not follow the model and behave properly. That's okay. Kids behave the same way and it's all part of the joy of living. The result is that instabilities might develop with severe vibrations or even wild behavior. (This sounds more like my family by the minute.) With the kids, we can experiment with cutting down on the sugar. With robots, we can consider taking two actions:

■ Perform the actions mentioned earlier to get rid of severe ringing.

■ Look for design flaws in the mechanics and control system that would make it more complex than the second-order system we're trying for. Look for places energy might be stored that we didn't expect. Change the design to compensate for it.

What happens when we take a second-order system and try to put it in a closed-loop feedback system? Well, consider the following closed-loop feedback control system (see Figure 2-19).

Let's assume the actuator is a second-order system such as the one we have studied. As we've seen, it will not react immediately to a step input function. It goes through some delay, a rise time, and then a settling time. Suppose we wildly put inputs into the

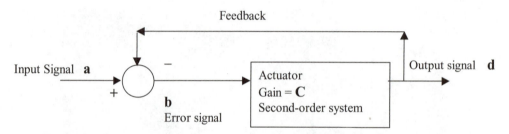

FIGURE 2-19   A second-order system used as the acutator in a closed-loop

input signal. Since the actuator cannot respond right away, output signal d would not change right away. The error signal b would reflect our wild inputs. The actuator input would see a wildly fluctuating input as well. If our input signals fluctuated somewhere near the natural frequency, ω, of the sytem, the output might actually ring out of phase with the input signal. This is exactly what happens when we oversteer a car. A car's suspension can be modeled as a second order system where:

- The mass is represented by the car itself.
- The springs are in the suspension.
- The damping friction is in the shock absorbers.

If we're driving a car and swing the wheel back and forth at just the wrong frequency, the car will weave back and forth opposite the way we're steering and go out of control.

Here's an example where a second-order system is overcompensated by a human feedback control system. Although most cars are well designed, little can prevent us from operating them in a dangerous manner. For whatever reason, this flaw in the design of cars is left in. What is needed is a filter at the steering wheel that prevents the driver from making input that the car cannot execute. A good driver will not oversteer and does so by not jerking the wheel around too rapidly. In effect, a good driver filters his actions to eliminate high-frequency inputs. This prevents the car from going out of control. You can do the exact same thing with your control system by putting a high-frequency filter on the input, ideally one that will attenuate input signals of a frequency higher than $\omega/2$. Since the the construction of filters is an art unto itself, it's left to the reader to study the technology and implement the design. Now let's tackle the third goal.

HOW TO ALTER THE ROBOT'S DESIGN PARAMETERS

We have already seen that altering ω and δ can substantially change the performance of the robot. Further, altering these parameters offers a reliable way to change just one type of behavior at a time without significantly disturbing the other behaviors. For instance, altering δ changes just the overshoot, with minimal changes to the rise time. Altering ω changes just the ringing frequency with minimal changes to the overshoot. Here's how to alter ω and δ:

- Altering ω
  - We know that $1/\omega^2 = m/K$.
  - $\omega = (K/m)^{0.5}$
  - To change ω, change K or m or both. We can change K by putting a different spring in. A stiffer spring has a higher value of K. We can change m by altering the mass of the robot.
  - Beware!
    - We know that $2 \times \delta/\omega = B/K$.
    - If we change ω or K, then we must change B if we want to hold δ constant.
- Altering δ
  - We know that $2 \times \delta/\omega = B/K$.
  - Given ω is held constant, in order to change δ, alter B if possible. Only alter K if we must.
  - Beware!
    - We know that $1/\omega^2 = m/K$.
    - If we change K, then change m to hold ω constant.
  - Most of us are familiar with a particular way of altering δ. Many older or used cars will exhibit a very bouncy suspension. When driven over a bumpy road, the car will bounce along and be difficult to control. The wheels will often leave the ground as the car bounces. Most experienced drivers will realize that the car needs new shock absorbers. But what exactly is happening here? The mass m of the car is not changing. The springs (spring constant K), installed at the factory near each wheel, have not changed. The shock absorbers have simply worn out. The shock absorbers look like tubes, about the size of a toddler's baseball bat, and are generally found inside the coil spring of each wheel. These shock absorbers are filled with a viscous fluid and provide a resistance to motion as the tires bounce over potholes. They exhibit a fluid friction coefficient of B. Unfortunately, the shock absorbers can develop internal leaks and the value of B decreases. When this happens, the overshoot of the second-order system becomes too great, and the wheels start to leave the ground. Replacing the shocks restores the original value of B and brings the overshoot back to the design levels. Bigger cars have more mass, bigger springs, and generally have larger shocks. Here is a PDF file and a web site dealing with the management of shock: www.lordmed.com/docs/ia_CATALOG.pdf

Let's tackle the fourth and final goal.

HOW TO GET OPTIMUM PERFORMANCE FROM THE ROBOT

The requirements for a second-order system might vary all over the place. We might need a fast rise time; we might need a quiet system that does not oscillate much; we might need to minimize mass or another design parameter. Don't forget that ω and δ are parameters derived from m, K, and B. We might be stuck with one or more of these five parameters and have to live with them. For example, the mass m might be set by the payload, the spring constant K might be inherent in the suspension, and the friction B might be set by the environment.

In many systems, the requirements are often at odds with one another and compromises must be struck. In such a design, it is often difficult to figure out what to do next. So here's a fairly safe bet. Take a close look at Figure 2-16. It shows four curves, including the lowest curve at a damping figure of 0.99. A second-order system with a damping constant near 1 is called "critically damped" (see Figure 2-20). The system rises directly to the level of 1. No overshoot or undershoot takes place. True, the rise time is nothing to marvel about, but the system is very stable and quiet. Designing a system to be critically damped is a good choice if no other definable target exists for its performance. It tends to be a very safe bet. In practice, it makes sense to back off from a damping constant of 1 a little bit, since an overly damped system is a little sluggish. If you can afford some overshoot, consider a damping constant between .5 and .9.

Notes on Robot Design

There are a number of other considerations to take into account when designing a robot. I've listed them here in no particular order. These are just tricks of the trade I've picked up over the years.

DESIGN HEADROOM

Cars offer great examples of second-order system designs. A car designer might be called upon to design a light car with a smooth ride. Ordinarily, a light car will bounce around quite a bit simply because it's smaller. Carrying this vision to an extreme, consider a car so small it has to drive down into a pothole before it can drive up the other side and get out of it. Certainly, a lighter car will suffer from road bumps more than a heavier car, but there is more to it than this. When a car goes over a pothole, the springs and suspension attempt to absorb the impact and shield the passengers from the jolt. But if the springs reach the end of their travel (as they would with a deep pothole), they

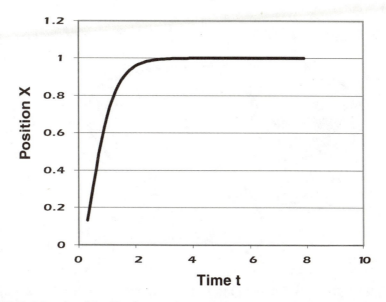

FIGURE 2-20 **A critically damped second-order control system is sometimes considered optimal.**

become nonlinear. In this situation, the second-order model breaks down, the spring constant becomes quite large for a while, and all bumps are transmitted directly to the passengers and the rest of the car. That's how you bend the rims, ruin the alignment, and get a neck cramp! It is up to us, as designers, to make sure the second-order system has enough headroom to avoid these problems. If your robot is to carry eggs home from the chicken coop, make sure the suspension is a good one (see Figure 2-21).

NONLINEAR CONTROL ELEMENTS

Thus far in our calculations and mathematics, we've assumed that all control elements behave in a linear fashion. Very roughly defined, this assumes a smooth, continuous action with no jerky motions. Bringing in a definition from calculus, this linear motion is characterized by curves with finite derivatives. Figure 2-22 shows a continuous curve and a discontinuous curve. Picture for the moment sending your robot over the terrain described by each curve and it will be easy to visualize why we should be considering nonlinear control elements in this discussion. We must be prepared to deal with such matters because most robots have some nonlinear elements somewhere within the design. Often, these elements are inherent in the mechanics or creep into the control system when we least expect it (see Figure 2-22).

FIGURE 2-21   This robot has an insufficient dynamic range in its shock-absorbing suspension.

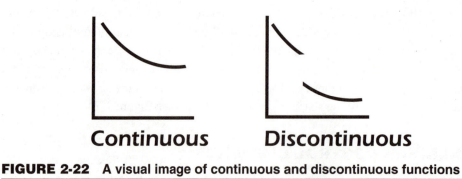

Continuous   **Discontinuous**

FIGURE 2-22   A visual image of continuous and discontinuous functions

Consider the case of an actuator or sensors that are either off or on. These are familiar to you already:

- **Thermostats**   The furnace in most houses cannot be operated halfway. The burners do not have a medium setting like a stove. Either the heater is all the way on or the heater is completely off. The thermostat represents the sensor feedback control input signal. It turns the heat all the way on until the temperature at the thermostat goes over the temperature setting. Then it turns the heat all the way off until the temperature falls below the temperature setting. It's expensive and inefficient (in terms of combustion) to ignite a furnace, and it's best if it runs for a while once it is ignited. The net result is that the temperature in the room doesn't stay at a single temperature. Instead, it cycles up and down a degree or two around the setting on the dial. This action, taken by many control systems, is called *hunting*. We'll talk about hunting shortly (see Figure 2-24).

This hunting action by the heating system is just fine in the design of the thermostat. Humans generally cannot sense, nor are they bothered by, the fluctuations of temperature about the set point. But consider a light dimmer. If the dimmer turned the lights on and off five times a second, reading would be rather difficult. Instead, dimmers turn the light on and off around 60 times a second so the human eye cannot sense the fluctuations. When you design a system that will have hunting in the output, be sure you know the requirements.

■ **Mechanical wracking**   Many mechanical systems have loose parts in them that will slip and then catch. In the model second-order system, consider what happens if the weight is mounted to the spring with a loose bolt. As the weight shifts direction, the bolt comes loose for a while and then catches again. The spring constant actually varies abruptly with time, and the smooth response of the system is disrupted.

You can model the robot's performance by considering that the model system will behave in two different ways. While the bolt is caught, the spring constant is per design. While the bolt is loose, the spring constant is near 0. If such a mathematical model is too difficult to chart, you can take the following shortcut. Just figure on adding the mechanical wracking distance (the distance the weight moves unconstrained by the bolt) to the overshoot and undershoot. This will make a good first estimate of its behavior. In practice, try to minimize the mechanical instabilities in the robot.

■ **Digital actuators**   Many other actuators and sensors tend to be digital. Consider a solenoid. It's basically an electromagnet pulling an iron slug into the center of the magnet. It's either off or on. The iron slug provides the pull on the second-order system when the electromagnet is activated (see Figure 2-23).

Effectively, our model of the second-order system is good for predicting the system's behavior since the solenoid behaves like a step input.

HUNTING

We've seen in the case of the thermostatic heating control system that the output of the system will hunt, effectively cycling above and below the temperature set point without ever settling in on the final value (see Figure 2-24).

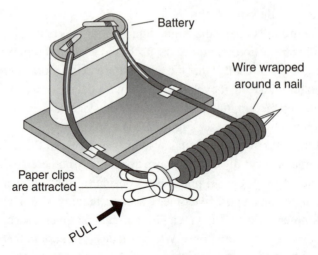

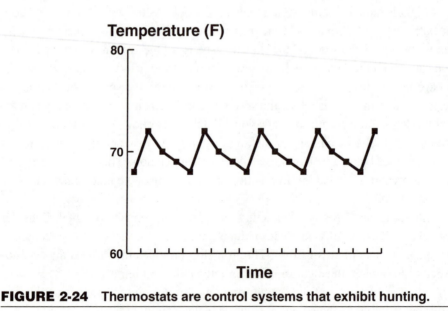

FIGURE 2-23   **Electromagnets exert pull inside relays, soleoids, and electric motors.**

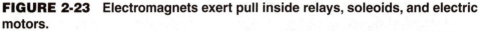

FIGURE 2-24   **Thermostats are control systems that exhibit hunting.**

In linear control systems with a great deal of power and some weaknesses in the high-frequency response, the output response will actually have a hunting sine wave on it. This disturbance can be quite annoying, much like the buzz in a stereo system. It's not unlikely that the oscillations would be at ω unless governed by a nonlinear element in the system (see Figure 2-25).

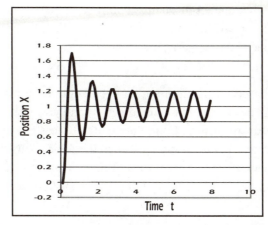

FIGURE 2-25 **A control servo system exhibiting unwanted sine-wave hunting**

Think for a minute how upsetting it would be if the elevator door opened and the height of the elevator oscillated up and down while you were trying to get off! In many systems, hunting is not acceptable. Hunting behavior can be avoided by refraining from using any nonlinear elements:

- Digital actuators that are on-off (like a solenoid) introduce nonlinear motion into a system.
- Don't use digital sensors that report only on and off. The sensors that turn on night lights are like this. They do not bring the lights on slowly as it gets dark.
- Avoid mechanical wracking. The mechanical parts of the robot may make sudden moves if all the bolts are not tight. The control system cannot compensate for this very well.
- Decrease ω. Often, if we decrease the frequency response of the system, we can avoid oscillations. Of course, this comes at the expense of slower performance.
- Add a hysteresis element to the control system; such an element is defined as "a retardation of an effect when the forces acting upon a body are changed." The common way to look at a hysteresis element is that it behaves differently depending on the direction. We are including here a few nonlinear control system elements that we can make a case for grouping with the hysteresis topic. Here are some examples of hysteresis elements:
  - A friction block that drags more easily one direction than the other.
  - A spring system that puts two springs into service when moving one way, but releases one spring when moving the other way.

■ An object with a ratchet mechanism on it so it moves one tick mark easily in one direction but will not move one tick mark the other way unless it's being forced to move two tick marks that way. Such a system is great for keeping the object still when it comes close to equilibrium (see Figure 2-26).

■ Gain changes based on position are another example. Elevators typically have powerful motors pulling them up and down when they are between floors. When they get very near the desired floor, they switch to less powerful motors to make the final adjustment before stopping. When the door opens, they may even turn off the motors completely. These sorts of gain changes make it much easier to avoid hunting in the final position of the control system (see Figure 2-27).

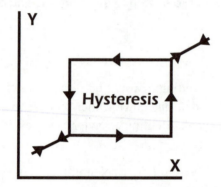

FIGURE 2-26 Mechanical (or electrical) hysteresis prevents symmentrical movement.

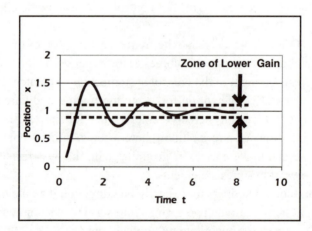

FIGURE 2-27 Control system gain can be decreased near equilibrium.

A CAUTION

So far, we've been talking about robot control systems in a very abstract way. The equations show very nicely that our mathematics will cleanly control the position of our robot in a very predictable manner. Further, we can smugly make minor parametric changes in the equation and our robot will blissfully change his ways to suit our best hopes for his behavior.

Well, it's very easy to get lost in such a mathematically perfect world. Those of us who have had kids are well acquainted with a higher law than math called Murphy's Law. Visit www.murphys-laws.com for the surprising history of Murphy's Law on the variants thereof that apply to technology. I had long suspected that such wisdom would be biblical in its origin, but it came into being in 1949.

Murphy's Law, as commonly quoted, states "Anything that can go wrong will go wrong." All along, we have been plotting and scheming to build and control a second-order control system. We've got that pretty well down. The trouble is our model will never exactly fit the real-world robot we're building. We have a mathematical control system that will control a single variable, such as our robot's position, to ever-exacting precision. However, this will not be the only requirement we will have to satisfy. We have ignored other unstated requirements along the way. To satisfy these other requirements, we may have to change the behavior of our simple control system, or we may have to put in even more controls. The following section on multivariable control systems speaks to this issue somewhat. Here's a few other requirements that are liable to crop up:

- **Speed**   Great, we've designed our position control system so our robot will move to where it belongs. But what about speed traps? Velocity is the first derivative of position. In the parlance of the variables we have been using, $v = dx/dt$. We really haven't worried about speed at all so far. Clearly, it is partially related to the rise time of the position variable. The quicker the control system can react to changes in position, the faster it is likely to go. But there will be various restrictions on speed:
  - **Safety**   Sometimes it's just not safe to have a robot moving around at higher speeds.
  - **Power**   Sometimes it's wasteful to go too fast. Some motors and actuators are not as efficient at top speed.
  - **Maneuvering**   Some robots don't corner well. It can be advisable to slow down on the curves.
- **Acceleration**   Fine, we've designed our velocity control system so our robot will not speed or be a hazard. But how fast can we punch the accelerator? Acceleration is the first derivative of velocity and the second derivative of position. In the parlance of the variables we have been using,

$$A = dv/dt = d^2x/dt^2$$

We really haven't worried about acceleration at all so far. But various restrictions on acceleration will take place:

- **Traction**   Wheels, if we use them, can only accelerate the robot a certain amount. Beyond the traction that the wheels provide, the robot will burn rubber!
- **Balance**   The robot might pop a wheelie.
- **Mechanical stress**   Acceleration imposes force on all the parts of the robot. The robot might rip off a vital part if it accelerates too fast. More on this later.
- **Mechanical wracking**   The robot will change shape as it accelerates. This happens in loose joints and connections. More on this later.

So with all these variables to control at one time, what do we do?

Multivariable Control Systems

Up to this point, we've been trying to build a control system for the robot that could serve to maintain a single variable, such as position. We should recognize that the mathematics of the control system are very general and apply just as well to robots that want to control other single variables like speed or acceleration. Although cruise control systems are very complex, they are simply control systems that regulate speed to suit the driver's needs.

But what happens if we want to control two or more variables simultaneously? Suppose we want the robot to follow a black line *and* move at a safe speed. Control of both position (relative to the black line) and velocity (so the robot does not veer too far off course during high-speed turns) puts us in the position of controlling two variables at the same time. How do we do this? (See Figure 2-28.)

One solution is to put two separate control systems into the robot. One system will control the position relative to the black line. The other control system will make sure the robot moves at the appropriate speed. Such a control system is inherently a distributed control system such as the ones we discussed earlier. Cars do, in fact, have multiple computers handling these tasks. Each control system has its own set of issues that we have discussed, such as steady state error, overshoot, ringing, and settling time. However, as we discussed in the section on distributed control systems, things can become complex very rapidly. Here's some points to consider:

- Wouldn't it make sense to slow the robot down if it is very far off the black line?
- Would it be a good idea to speed up if the robot has been on course for quite a while?

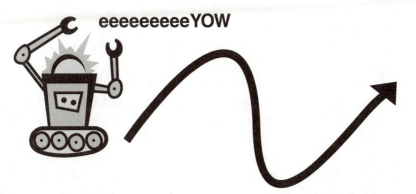

eeeeeeeeeYOW

FIGURE 2-28   It's hard to control two variables at the same time (such as speed and direction).

- What do we do if one of the control systems determines that it is hopelessly out of control? If it loses track of the black line, should it slow down?
- If the robot is moving very rapidly, does it need to look farther ahead for bends in the black line?

All the scenarios argue for sending information back and forth between the two control systems. Further, the ways in which they interact can become very complex. At some point, if more and more control systems are added to the robot, the following can occur:

- Multiple control systems get expensive.
- Communication between the control systems can get expensive and slow things down. In the worst case, communication errors can occur.
- Interactions between the control systems can get unpredictable. In the worst cases, instabilities can arise. These instabilities can take the form of unexpected delays or thrashing. Thrashing arises when two control systems disagree and fight over the control of parts of the system. Each control system sees the actions of the other as creating an error.
- Designs can become very complex to accommodate all cases.
- Designs can become difficult to maintain. As one control system is changed, other control systems may cease to function. Retesting the robot becomes a large task.

Many years ago, in the primordial soup of engineering history, engineers began to consider control systems that had more than one variable. We need only look at old drawings of steam engines to appreciate this. They had to regulate speed, pressure, temperature, and several other variables all at the same time. The general approach back then was to put multiple mechanical control systems in with interlocks as needed. Failure meant explosion!

The speed governor in Figure 2-29 is a great example of a mechanical engineer used to solve a control system problem. It regulates the speed of an engine. As the engine speed increases, the two metal globes spin around the vertical shaft. Since the outward centrifugal force increases, the globes start to move outward, pulling on the diagonal struts. The diagonal struts, if pulled hard enough, will pull up the base and release some steam pressure. This keeps the engine from going too fast. It's a good example of a separate control system for velocity. School buses still have such mechanisms on their engines if you look carefully. But better ask permission before snooping around!

A nice example of a governor design can be found at www.usgennet.org/usa/topic/ steam/governor.html. A few years later, engineers began to think about centralizing control systems. Computer electronics facilitated this transition since all the information could easily be gathered in one place and manipulated. The engineers cast about for a way to control multiple variables at the same time and raised several key questions:

- How would a multiple variable system be designed? What framework would it have?
- How many variables could be controlled at the same time?

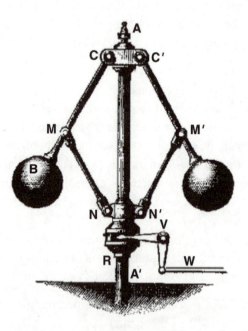

FIGURE 2-29 The speed governor is a venerable mechanical feedback control system.

- What equivalent exists for a "steady state error" in a system with multiple variables?
- How do we evaluate the relative state of the control system? How far is it from the optimal control state? What is the error signal?
- How can we alter the design of the system to affect its performance?

Let's look at the first question.

HOW WILL WE DESIGN THE MULTIPLE VARIABLE SYSTEM? WHAT FRAMEWORK WILL IT HAVE?

Let's assume for simplicity's sake that we are trying to design a control system to control just two variables at the same time: X1 and X2 (perhaps position and velocity). The following discussion can be generalized to *n* variables (X1, X2, X3 . . . Xn) on the reader's own time. We can call the combination of the variables X1 and X2, the vector X.

Let's assume that the desired state of the two control variables is as follows:

- X1 = X1d
- X2 = X2d

We can call the desired state of vector X, the vector Xd.

If computers are used in the control system, the computer periodically finds a way to change X based on the value of Xd. In such a control system, we speak of computations executed at periodic, sequential times labelled t − 1, t, t + 1, and so on. We use the following notation:

- X(t − 1) shows the values of X at the previous computation time.
- X(t) shows the values of X at the present computation time.
- X(t + 1) shows the values of X at the next computation time.

Similarly, Xd(t) represents the time series of values for Xd.

To compute the next value of X1, for instance, the computer will look at the previous and present values of both X1 and X2 and determine which way to change X1 in an incremental way. The same computation is done for X2. Done properly, X1 and X2 will slowly track the desired values. But how do we go about finding the iteration?

Iteration is a process of repeating computations in a periodic manner toward some particular goal. Usually, an iteration equation governs the process of iteration. The following is a general-purpose iteration equation that is often used in robots. X(t) is computed by iteration by taking values at time t and iterating to the next value at time t + 1:

$$X(t + 1) = X(t) - S(t) \times (d\, C(X(t))/d\, X(t))$$

In the equation, S(t) is a vector of step sizes that might change with time but can be fixed. This vector could contain, in our example, two fixed step size values, each roughly proportional to 5 percent of the average size of X1 and X2. An alternate method could have the vector contain two varying step size values, each roughly proportional to 5 percent of X1 and X2's present values. The point is X1 and X2 will change gradually in a particular direction in order to satisfy control system requirements. If the cost function C(X(t)) shows that X1 must increase, then the time iteration of the equation will bump X1 up by the step size. If the cost function shows that X2 must decrease, then the time iteration of the equation will bump X2 down by the step size.

C(X(t)), a vector of cost functions based on X(t), is yet to be defined. The cost function is a measure of the "pain" the control system is experiencing because the values (past and present) of X(t) do not match the desired values of Xd(t). We use the derivative (d C(X(t))/d X(t)) because we want the corrective step size

■ To be larger if the cost (pain) is mounting rapidly as X(t) changes the wrong way. Thus, we must take more drastic corrective action.

■ To be smaller if the cost (pain) is not mounting rapidly as X(t) changes the wrong way. We are near the desired operation area and are not in pain, so why move much?

Such an iteration equation can be used as a solution for robotic control. But what's missing is the cost function. The proper choice of a cost function really determines the behavior of the robot. Much of modern work on control systems revolves around the choice of the cost function and how it is used during iteration.

One very popular framework to give the control system is the least squares framework, discovered by Legendre and Gauss in the early nineteenth century (see Figure 2-30). Termed the *least mean square* (LSM) algorithm, it sets the cost function C(X(t)) proportional to the sum of the squares of the errors in each element of the vector:

$$C(X(t)) = k \times \Sigma_n (X(t) - Xd(t))^2$$

where k is an arbitrary scaling constant.

In our specific example, we could set the cost function to the sum of the squares of the errors:

$$C(X(t)) = 0.5 \times ((X1(t) - X1d(t))^2 + ((X2(t) - X2d(t))^2)$$

Differentiating by X1 and X2, we get the two elements of (d C(X(t))/d X(t)):

$$d\, C(X1(t))/d\, X1(t) = X1(t) - X1d(t)$$
$$d\, C(X2(t))/d\, X2(t) = X2(t) - X2d(t)$$

FIGURE 2-30 Gauss and Legendre

The cost function increases in magnitude as the square of the errors. The step size, used to recover from errors, then increases linearly proportional to the error. Specifically then, since

$$X(t + 1) = X(t) - S(t) \times (d\,C(X(t))/d\,X(t))$$

we have the two elements iterated as follows:

$$X1(t + 1) = X1(t) - S1(t) \times (X1(t) - X1d(t))$$
$$X2(t + 1) = X2(t) - S2(t) \times (X2(t) - X2d(t))$$

If we were to set step sizes S1(t) = S2(t) = 0.1, then

$$X1(t + 1) = 0.9 \times X1(t) + 0.1 \times X1d(t)$$
$$X2(t + 1) = 0.9 \times X2(t) + 0.1 \times X2d(t)$$

Thus, X1 and X2 slowly seek the values of X1d and X2d. Also, X(t) slowly seeks the value of Xd(t).

Before we look at cost functions other than LMS, let's finish answering some of the other questions we posed earlier.

HOW MANY VARIABLES CAN BE
CONTROLLED AT THE SAME TIME?

Practically speaking, the LMS algorithm can handle an arbitrary number of simultaneous variables. However, as the number of variables increases, the danger of interactions increases drastically. The primary danger is that unknown interactions between the variables will throw off the calculations and destabilize the control system. This often shows up in the math if the variables are not completely independent. In our example, the derivative of X1 with respect to X2 may not truly be zero, or vice versa. This would greatly compromise the stability of the stepping iterations. As a general rule, try not to use a single control system to handle too many variables at the same time. Two to four variables is a good place to stop.

WHAT IS THE EQUIVALENT FOR
STEADY STATE ERROR WHEN USING
MULTIPLE VARIABLES?

First of all, where multiple variables exist, be aware it's entirely possible the system will never come to a steady state. However, it is possible for the digital calculations to settle into a completely stable and quiet solution. Such a solution would have X(t) stable and equal to Xd(t).

However, with certain minimal step sizes, it may not be possible to converge on a quiet solution. Think for a minute of a system at 9, seeking 10, with a back and forth minimal step size of 2. The system will likely bounce back and forth from 9 to 11 and back to 9 forever. A carefully designed control algorithm can avoid such a problem, but we leave it up to the reader to work this out.

HOW DO YOU EVALUATE THE RELATIVE STATE
OF THE CONTROL SYSTEM? HOW FAR IS IT
FROM THE OPTIMAL CONTROL STATE?
WHAT IS THE ERROR SIGNAL?

For an LMS system, you can track the size of the cost function. All the terms in the sum are positive, squared numbers. The magnitude can be used as a measure of the state of the system. We clearly want it to be small. Further, the first derivative of the cost function should be quiet. The relative noise level of the cost function is a measure of the volatility of the system and it can be used to indicate disruptions at the inputs of the system.

HOW CAN WE ALTER THE DESIGN OF THE SYSTEM TO AFFECT ITS PERFORMANCE?

An LMS algorithm is relatively straightforward for the following reasons:

- We can keep the step sizes in the vector S(t) as constants. If the step sizes vary between 0 and 1, the system response speed varies from glacial to jack rabbit. We must recognize that jack-rabbit control systems have too high a frequency and are vulnerable to overshoot, ringing, and instabilities. A good bet is to get your robot working first and then back down the values of S(t).

- We can alter the step sizes in the vector S(t) to keep the rest state of the system quiet. The way in which this is done must be chosen with great care to avoid adding noise to the system. One good bet is to decrease the step sizes as the system starts to quiet down, and increase the step sizes (within reason) as the system begins to get noisy and active.

- We can alter the step sizes in such a way that they are always a power of 2 (like 1/8, 1/4, 1/2, 2, 4, 8, 16, and so on). Multiplying (or dividing) by a power of 2 only requires a simple shift operation in binary arithmetic. Restricting the step sizes to such values can make LMS computations much simpler for smaller microcomputers to execute.

- We can set the step size to 0 when the cost function is small enough. This will prevent thrashing around near the optimal solution. Such thrashing around can be caused by input noise and by minor arithmetic effects. Picture an elevator opening its doors. The passengers are no longer interested in getting exactly to floor level as long as it's close enough. The passengers would be truly upset if the elevator control system was still moving up and down a tiny bit trying to get it just right. Instead, elevator control systems stop all action when the doors open. We can achieve a similar effect by setting the step size to 0. We will look at other safety considerations later.

NON-LMS COST FUNCTIONS

A control algorithm, like LMS, has behavioral characteristics that will affect how our robot will behave:

- LMS control systems tend to react slower to inputs. This usually means they have slower reaction times.
- LMS control systems are more stable in the face of noise on the inputs.
- The math is not difficult and does not consume valuable computer resources.

Other cost functions beyond LMS are available. LMS still requires multiplication, which can eat up computer time and resources. LMS multiplies the step size by the differential error $(X1 - X1d)$ to get the iteration step size. This can be approached in other ways:

- Use just the sign of $(X1 - X1d)$, not the magnitude. The sign simply indicates which way X1 is off. The entire step size is then simply added or subtracted from X1 to iterate to the next value. This makes the iteration step a simple addition or subtraction and avoids the multiplication. This can be of particular value if we choose to use a small microcomputer that has no multiplier.
- Use the relative size of $(X1 - X1d)$ to pick the step size from a table of step sizes. This can work well and also avoids multiplication. It can converge faster when the cost function is large and can remain fairly quiet about the optimal solution. Care must be taken when switching gears in an arbitrary manner like this. Please reread the earlier "A Caution" section.

Multivariable systems have other peculiarities to worry about as well. Issues of stability, convergence, and speed of operation all must be addressed here:

- **Stability** As already discussed, if the step size is too large, the system may oscillate about the solution point in an unacceptable manner. Further, all the variables may not be able to reach an optimal solution at the same time. The system may remain noisy forever, even if the inputs stop moving.
- **Convergence** It's possible, in some situations, that the control system will not actually move to an acceptable solution:
    - **Finding a solution** Sometimes the starting position of the robot can affect whether it will move to the desired location or not. The control system always has a set of points beyond which it cannot recover. In the design and operation of our robot's control system, we must assure ourselves that the robot will not be asked to recover from such a situation. Note that we must determine what an acceptable solution is for the robot. Often, this involves some metric on the size of the cost function, but this can be done many different ways.
    - **Avoiding false solutions** Sometimes arithmetic systems will settle into a false solution. An example might be a robot looking for the highest hill, only to find a smaller hill nearby. If the control system must contend with a complex environment, this can happen easier than we might suspect. If the situation looks suspicious, consider putting some safety mechanism into the control system that will jar the robot out of a false solution if it gets stuck in one. Such

a "safety" system must be very well designed to make sure it does not create a false alarm and disrupt a perfectly good solution.

- **Speed of operation**   As with any robot control system, good performance is always expected. The speed of operation is almost always one of the criteria. If the step sizes are too small, it might take intolerably long to move to the proper solution. Choose the step size to optimize the robot's behavior in terms of speed and accuracy. Consider choosing the step size to best match the capability of the robot to move and maneuver. If the match is close, the results will be better in the form of smoother operation.

Now we need a bit of a reward for having slogged through so much "useful" math. It's time to dream a bit and talk about more esoteric matters that might not affect us today or tomorrow but are important anyway.

Time

A little ways back in this book, we talked about the fact that the earth cannot be counted on to be a stable reference point for our robot. As a practical point, it truly is stable enough in every case I've ever seen, so I'm content not to worry about the earth.

But along comes Albert Einstein to throw us another curve ball (see Figure 2-31). It turns out that we cannot count on time itself to be unvarying in our calculations. However, if the robot is puttering around at a slow speed and stays away from black holes, we can probably ignore the considerations that follow. If the robot will be moving at high speeds relative to the earth, then Einstein's calculations come into play.

In the very early 1900s, Einstein came up with the special theory of relativity, which holds that time does not always run at the same rate. If two bodies are moving with respect to one another, they will experience time running at two different rates. The effect does not become serious until the speeds are high. But even the astronauts circling the earth have to take relativisitic time into account or their orbital calculations will be off. The following URLs show some of the calculations involved in the theory. It was a Polish mathematician Minkowski who provided the math that eluded Einstein.

- www.astro.ucla.edu/~wright/relatvty.htm
- www.physics.syr.edu/courses/modules/LIGHTCONE/twins.html

Time varies roughly as $1/\mathrm{sqrt}\,(1 - (v/c)^2)$, where v is the relative velocity of the object and c is the speed of light. Using this formula, plugging in an orbital speed of

FIGURE 2-31 Einstein

roughly 8,800 meters per second, and given the speed of light at roughly 300,000,000 meters per second, we get a time dialation for an orbiting spacecraft of

$$1/sqrt\,(1\ -\ (8800/300{,}000{,}000)^2\,)\ =$$
$$1/sqrt\,(1\ -\ 0.00000000086)\ =$$
$$1.0000000004$$

So, consider the Soviet cosmonaut who spent 458 days in space (the record) (for a total of $458 \times 24 \times 60 \times 60 = 39{,}571{,}000$ seconds). Ignoring all the other motions of the spacecraft other than the orbital speed, the cosmonaut's time dialated $39{,}571{,}000 \times 1.0000000004 = 39{,}571{,}000.017$ seconds.

Thus, after over a year in orbit, a time change of 17 milliseconds has occurred for the cosmonaut. That's not very much, but at an orbital speed of 8,800 meters per second, the cosmonaut would be off by 150 meters (8800×0.017). That's not very far in terms of the earth's expanse, but a big error while you're trying to dock! Orbital planners do take relativistic effects into account in planning orbits and interplanetary missions.

Space

Well, if it's not bad enough having to worry about just what time is, Einstein threw another monkeywrench into our collective thinking. The General Theory of Relativity holds that the fabric of space itself isn't just a series of straight perpendicular lines like some street pattern, but rather it's curved and changing! He came up with this theory using a truly beautiful "thought experiment." Instead of working in a lab, Einstein sat down and pictured the experiment in his head. Here's how his thinking went.

Suppose we are sitting in a room in far outerspace where no gravity exists. Two holes are in the wall, one to the left and another to the right. A beam of light comes in one wall and out through the other. It does not take long for the beam of light to cross the room at light speed. Light travels one foot per a billionth of a second (see Figure 2-32).

Now, if you accelerate the room upward at 32 feet/second/second (1 G of gravity), when the next beam of light comes through the first hole, it won't make it out through the second hole (which has now moved). From our standpoint sitting the the room, the light beam curves after it enters the room and hits the wall too low (see Figure 2-33).

Now suppose instead of acceleration, we put the earth immediately under the room. From our standpoint sitting in the room, we could not tell the difference. We still experience 1 G of accelerative force under us. The beam of light comes in the first hole and still bends down to hit the wall below the second hole (refer to Figure 2-33).

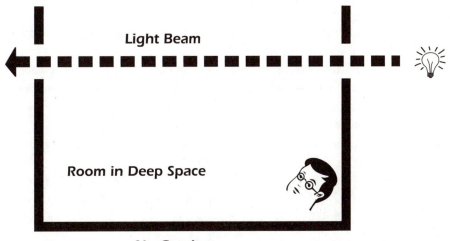

FIGURE 2-32 Einstein's thought-experiment: Light moves straight in the absence of gravity.

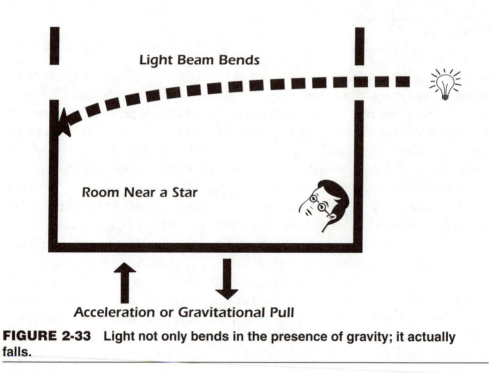

Light Beam Bends

Room Near a Star

Acceleration or Gravitational Pull

FIGURE 2-33 Light not only bends in the presence of gravity; it actually falls.

Gravity is this bending light. But if we maintain that light must travel in a straight line at a constant speed, then we must conclude that gravity bends space itself. The very existence of matter, which engenders gravitational force, bends our fabric of space.

Seems simple enough, right? Lest you worry about your warped existence, please be assured that the bending of space is quite small and can be ignored in most of our everyday existence.

Around the First World War, some astronomers decided to put Einstein's General Theory of Relativity to a test. They observed some known stars during a solar eclipse. Sure enough, stars emerged from behind the sun and moon earlier than they were supposed to. The stars' light was coming from behind the sun (where the astronomers should not have been able to see it), bending around the sun's gravity and appearing before they were supposed to. Further, the amount of the observed bending closely matched Einstein's theoretical calculations. This was a revelation in the sciences and confirmed Einstein's major discovery. It was a beautiful piece of work (see Figure 2-34).

A few years after that, scientists found three stars in a row, with the outer two appearing identical. It turns out that the light from one star was being bent around an intervening star, so both images appeared to us on Earth. This was another manifestation of gravity bending light and has been called a *gravitational lens*. Since starlight can bend

FIGURE 2-34 A gravitational lens. The path of light defines straight lines, so gravity bends space.

around an intervening star in any direction (360 degrees), gravitational lenses often provide an image of a star as a ring or arc of light. Some nice examples of gravitational lenses can be found at www.iam.ubc.ca/~newbury/lenses/glgallery.html.

The web page at http://imagine.gsfc.nasa.gov/docs/features/news/07nov97.html has reported an extreme case of this effect as "a black hole that is literally dragging space and time around itself as it rotates . . . [in] an effect called frame dragging."

COMPUTER HARDWARE

Before getting into the nuts and bolts of choosing the computer hardware to include in the robot, let's take a step back. What are the reasons for putting a computer inside the robot? Even experienced engineers choke on this question. It seems, after all, to be a natural decision. Yet when we look at any one particular reason, there always seems to be yet another underlying reason behind it. At the beginning of any one phase of the robot project, it makes sense to analyze the options. Often, a better solution is at hand. Let's look at a nontechnical example.

You and your friend are in an open field and are confronted by a hungry lion (see Figure 3-1). The lion starts to charge and it is clear you must run. What should your immediate goal be? Some say, "Outrun the lion." Others say, "Outrun your friend."

Clearly, it can be difficult to think in stressful situations. If we have time to think, a better solution can usually be found that will save us much time, effort, and pain. Do not, however, get trapped in endless rounds of thinking and planning. This too is a good way to get eaten by the lions.

This survival scenario is a good example of how larger questions always reside above the immediate question. Did the second answer above make you smile? If so, why?

FIGURE 3-1 A hungry lion can be a problem.

So why use a computer at all? The bottom line is

- The project will cost less to complete.
- The robot will be a better one.
- The design can be finished sooner.

Let's look at where these savings accrue. Every project has costs in terms of time and money:

- **Cost** What types of cost exist?
  - Direct cash outlay for equipment, parts, and tools.
  - Tying up scarce resources. Sometimes projects consume resources that cannot be replaced but are essentially free. An example would be the time of a key employee. If another project came along, the key employee would not be available.
- **Development time** The amount of time the development takes has various costs attached to it. If the schedule for a commercial robot project slips, a company can miss a large percentage of the potential profits. As soon as competitors come out with similar products, profits drop off quickly. The first few months of a product's lifetime are the most valuable. If the robot is not ready on time, the opportunity cost is lost. If a project schedule slips, real costs generally run up. Resources and personnel can also be tied up, causing a longer development time.
- **Risk of failure** Managers of robot projects often expend resources early in the schedule to defuse risks. As an example, consider a robot that must traverse diffi-

cult terrain. The designers may choose to build a couple of different drive trains and test them out before proceeding with the rest of the project. If a project has few risks, the final cost is likely to be lower. If the risk items become real problems, schedules often slip and costs run up.

The decision to use computer hardware in the robot design can decrease the cost of the project in various ways. The following section illustrates a few ways to make this a reality.

Leverage Existing Technology

"If I have seen further, it is by standing on the shoulders of giants."
Sir Isaac Newton (Figure 3-2), cited in *The Oxford Dictionary of Quotations*

Civilization advances on the strength of its history and knowledge. Humans are unique in that we store information outside our brains, in libaries and computers. The accumulated work of others can be brought to bare to solve our problems. In the case of computers, engineers have made their work available in the form of archived software and printed circuit hardware. Each can be rapidly and inexpensively reproduced for our use.

Computer hardware is available in various forms. We can purchase complete computers at stores, but these tend to be too bulky to fit into a robot. We can purchase

FIGURE 3-2   **Sir Isaac Newton**

printed circuit cards from distributors and place them inside the robot. We can also purchase computer chips from distributors and build our own printed circuit cards, a difficult proposition for the casual robot designer.

We can purchase complete computer systems on a card, which will accept our software and provide connectors for the signal lines we need to control the robot. This is often the most economical method of integrating computers into the design, unless large quantities of robots will be manufactured.

The companies that sell computers have invested millions of dollars to make their technology available for our use. We gain time, dollars, and reliability by sharing and taking advantage of their effort. Because the technology has been made so readily available to others, many third-party designs are also available for us to use, such as the following:

- **Third-party hardware** Most computers have connectors on them that enable us to use the "bus." We'll define the term later, but suffice it to say, a bus allows third-party companies to design hardware that will plug right in to the computer. Dozens of *printed circuit boards* (PCBs) and other conveniently packaged circuitry are available.

- **Third-party software** It's not unlikely that other companies have written software we can use. If the computer we choose is "special purpose" (to be defined later), then several companies have probably written software that takes advantage of the special features of the computer. We can purchase this software and use it in various ways:

  - **Freeware** Often an author of software will make it freely available for others to use. One can search for "freeware" on the Internet, qualified by words that describe the software needed. Sometimes the author will ask for attribution or have other requirements.

  - **Shareware** Shareware is much like freeware, except the author often requests payment if the shareware is used in a robot. One can search for shareware in the same manner as freeware and one should read the restrictions very carefully. Make copies of the author's requirements and save them if questions should arise later. Searching for shareware takes some time, but it can be a very valuable endeavor. If nothing else, it can tell us how difficult our software effort will be. If it's easy to write and valuable, somebody else will have written it already. If it's hard, nothing remotely close will be available in shareware. We can also discover shareware that comes close and, along with it, the authors who might be employed to modify it for our project.

  - **Licensing** Large software operating systems, tools, and application software usually have licensing requirements. Contact the company that sells the software directly for information.

Speeding Up Engineering

Using computers within the robot obviates the need for full and detailed planning.

Now I've done it! Of all people, I advocate planning as a time-saving effort that is well worth engaging in. The truth is, some projects are too difficult to plan all the way through in great detail. But if we can be reliably assured at the start that our computer will give us the flexibility and horsepower we need for unforeseen circumstances, we can proceed without full planning.

Putting a computer in the system brings the following benefits to the engineering schedule:

- The overall engineering effort can be partitioned. If we have more than one person working on the robot, the work can be divided and executed in parallel. One person can concentrate on the hardware while another person starts on the software. The hardware does not have to be finished before the software can start. The programmer can work on a board similar to the one in the robot.
- Changes in the specification of the robot can be made along the way with some confidence that the new requirements can be accommodated in just the software. It's much easier to change the software than to change a hardware design.
- The design can be changed as needed for future maintenance even after the robot is completed.

On the lighter side, one way to speed up engineering is to make a contest out of it. The following URLs show just how fast things can get done if we would just apply ourselves with diligence to an engineering problem:

- http://kennedyp.iccom.com/text/Playing_with_fire.txt
- http://home.att.net/~purduejacksonville/grill.html

Computer Architecture

Computers were designed to perform arithmetic calculations rapidly in a repeatable manner. There are many different ways a computer can be constructed and this section covers many of the different architectures that exist.

TYPES OF COMPUTERS

Let's assume, for the moment, that we've decided to put a computer into the robot. Although many general-purpose computers are available, it makes sense to take a look at the special-purpose computers first. It's likely we'll be choosing a general-purpose computer for the robot, but special-purpose computers can bring many advantages to the design. Before we take a close look at the architecture of the general-purpose computer, here is a quick tour of the basic architectures of some special-purpose computers.

Analog Computers

Webster's dictionary defines analog as "something similar to something else; a mechanism in which data is represented by continuously variable physical quantities." Analog computers are commonly perceived as a throwback to the early days of computing machinery. Even now, all electronic computers use analog electronic signals to support their calculations. General-purpose digital computers, however, restrict the analog electronic signals to just two voltage levels representing binary 1 and binary 0 in an effort to gain speed. Analog computers have no such voltage restrictions for signals. Instead, signals vary throughout the range of voltages that the analog computer electronics can support. A single analog signal can directly represent, for example, the speed of the wind from 0 to 255 mph. A general-purpose computer needs eight signals ($2^8 = 256$) to represent the same range of values for the wind.

Analog computers use analog electronics, such as operational amplifiers, to build circuits to simulate the behavior of complex systems. They are especially good at simulating systems that are governed by differential equations. The second-order control system described elsewhere in the book is a prime example. With just one operational amplifier, an analog computer can fully simulate the same curves and parametric controls we have already looked at. The front of an analog computer looks like a giant switchboard with lots of places to plug in wires.

To program an analog computer, an engineer uses patch wires to plug together the required building blocks. Knobs on the analog computer can be rotated to enter the values for the desired frequency and damping. The engineer starts the computer and a meter needle shows the resulting curve over the span of a couple of seconds of simulated time. In the example of our robot's second-order system, overshoot is evident if the meter needle goes too high before settling down. Ringing can be seen as the oscillation of the needle back and forth while it settles down.

Analog computers have dropped by the wayside for two basic reasons:

- A general-purpose computer can be programmed to simulate an analog computer, obviating the need for the analog hardware.
- General-purpose computers can be programmed in different ways to solve the same problems. Instead of simulating the analog computer (which simulates the real-world problem), a general-purpose computer can be programmed to simulate the real-world problem directly.

More information about analog computers can be found at www.science.uva.nl/faculteit/museum/AnalogComputers.html and at www.play-hookey.com/analog. The Analog Computer Museum, dedicated to the history of analog computers, is at http://dcoward.best.vwh.net/analog/.

Neural Networks

One of the finest computational engines known to exist is the human brain. It can solve most complex, real-world problems much faster than a general-purpose computer, albeit with less precision. Electronic computers are best suited to problems requiring arithmetic capability and blinding execution speed, such as forecasting the weather. But they are not good at solving problems requiring judgment or experience. The human brain has the experience and "wiring" to take on problems that it has never seen before and to solve them with speed and reliability. The parents of teenagers might argue with this last statement, but they have never tried to live with a teenage robot struggling with its computer's programming so it can survive puppy love. Be assured, parents would rather deal with a human teenager who, believe it or not, has amazing abilities compared to a computerized robot.

So what is a neural network? Ever since humans first grasped the structure and purpose of the human brain, they have dreamed of building an artificial brain. Many designs for such a brain have been put forth, including neural networks. First, let's look at the human brain.

Brain cells, called *neurons*, are connected together in a vast array of tissue within the brain. They communicate electronically with one another over neural connections called *synapses*. This allows neurons to exchange information with nearby neighbors. Neurons retain information (dubbed *memory*) chemically and electrically within the cell body (see Figure 3-3).

The memory of a specific spring day, for example, might be spread out over a vast array of neurons, which govern smell, sight, hearing, motion, and so on. The memory of the spring day is distributed throughout the brain. Memories can be imperfect and they can fade as individual neurons begin to lose their individual memory of the day. Memories are stored almost like a photo spread out over the fabric of the brain. Neurons might store more than one memory at the same time. This is why the remembrance of one thing, like a spring day, might evoke the memory of another experience, like the ice-cold water of a stream. A human, prodded to remember the spring day with the noise of a brook, would likely dredge up the memory of stepping into a noisy, icy brook. The fact that noise was in both memories ties the memories together. The human has learned to be suspicious of brooks on spring days; they might be icy.

Learning is something general-purpose computers are not good at. Some neural-network computers are designed to mimic the learning ability of the human brain. They

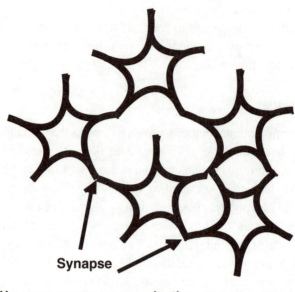

Synapse

FIGURE 3-3   **Human neurons communicating across synapses**

are exposed to a series of situations and gradually learn how to deal with them. Neural-network computers are generally designed with individual "neurons" that can communicate with one another, especially within their immediate vicinity. They are arranged in rows and banks of neurons; an example is shown in Figure 3-4.

The results of each layer are fed into a series of communication units that perform calculations and reroute information to other neurons. The flow of information is shown in Figure 3-4. A series of real-world events is fed into the inputs at the top; the neural net processes the inputs and generates responses out the bottom. The results are scored (by an experienced person) and the score is fed back into the neural network at the top. The network then readjusts its communication units so it will do better next time. Certainly, the network will do better the next time it sees the very same events fed into its inputs. But oddly enough, it often does better on new events it has never seen at its inputs before. As such, it is learning.

Neural networks can be built in many ways. One researcher took a silicon substrate (a slab used to build computer chips), hollowed out pits in the substrate, put neurons into the pits, and allowed the neurons to communicate by connecting synapses. Computer circuitry was etched in other areas of the substrate. The entire circuit ran on a combination of glucose and electricity.

Neural networks can be built from hardware (using computer chips) or they can be simulated in software. There have been many successful applications of neural network

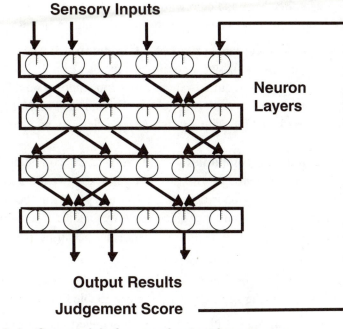

Sensory Inputs

Neuron Layers

Output Results

Judgement Score

FIGURE 3-4   One model of a neural network computer

software in systems that must develop "judgment." One application has been the prediction of credit card fraud. By exposing the neural network software to many credit card applications and then telling the network which customers defaulted later, the network is trained to scan new applications and reject those customers who might default later.

Here are some URLs for further study about neural networks:

- www.emsl.pnl.gov:2080/proj/neuron/neural/what.html
- http://vv.carleton.ca/~neil/neural/neuron.html
- www.cs.stir.ac.uk/~lss/NNIntro/InvSlides.html
- http://hem.hj.se/~de96klda/NeuralNetworks.htm

Special-Purpose Processors

The primary advantage of a computer is its blinding speed. It can execute many millions of instructions every second. But some tasks require the processing of truly massive amounts of information. These applications require the addition of even higher speed hardware to process the information. Such high-speed hardware is specifically designed to process the information at hand, but it can perform no other function. The high-speed hardware is integrated directly on the chip with the rest of the computer hardware. We can find special-purpose processors among the following supplier groups:

- *Application-specific integrated circuits* **(ASIC) vendors**   If we cannot find the specific special-purpose computer we desire, we can make one! Massive amounts of development dollars are required, so our robot application would have to have a really high sales volume to even consider this. *Advanced Risc Machine* (ARM) computer cores can be paired with special-purpose circuitry and put on individual ASICs.

- **Fabless semiconductor companies**   Many very small computer companies build special-purpose computers. Usually, they go to ASIC vendors to make their designs into chips, but they have done the work and spread out the costs among many customers. Find them in electronic design magazines and at conventions. Consider searching for them on the Internet using the special-purpose function as one of the keywords.

Many special-purpose functions have been integrated into computer circuits and brought to market. The following special functions are available from several suppliers:

- **Wireless communications**   Chips exist that can convert and convey *radio frequency* (RF) data signals directly into the computer circuit. These chips are used in pagers, phones, radios, *global positioning systems* (GPSs), RF identification tags, smart cards, and so on. If the robot application requires special-purpose computers with similar capabilities, consider looking at the suppliers in these markets. Be aware, however, that few of these chips are available in small quantities. They are also difficult to apply.

- **Internet communications**   Many computer chips are available with integrated *local area network* (LAN) interfaces that are used to connect to the Internet. Further, some of these computers have integral software stacks that can process the flow of Internet data in real time inside the chip. This sort of processing can greatly speed up a robot if its design requires a great deal of information flow over the *Internet Protocol* (IP).

- *Digital signal processing* **(DSP)**   DSP circuitry (to be defined shortly) is used to process information in ways most general-purpose processors cannot. Study the following DSP section. If a DSP is needed, consider
  - Texas Instruments' OMAP DSP processor at www.TI.com
  - Analog Devices at www.analog.com

- **Analog controllers**   Many special-purpose processors have analog circuitry right on the digital chip. One buzzword for this type of circuitry is *mixed signal*. Such a technology has several advantages, but the leading one is cost. If the chip can support all the requirements of our robot without further analog design effort, we can come out ahead. Consider Analog Devices' mixed signal family at www.analog.com/technology/dsp/mixedsignal/index.html.

- **Display systems**   Many robots require control panels or information displays. It is not difficult to integrate a *liquid crystal display* (LCD), even a large one, into a computer circuit these days. Many computer chips can support LCDs directly.

- **Low-power units**   The handheld *personal digital assistant* (PDA) market, along with phones and pagers, has spawned a whole series of computer chips that can operate on very low levels of voltage and power. If the power for our robot's computer system is a significant part of the power budget, then consider low-power computer systems. Many other techniques for saving power in computer systems can be used as well. We'll visit power control later in the book.

- **Game units**   It's a little-known fact, but most computers wind up in games. That's right. The sheer number of computers going into toys dwarfs the other practical uses. These are generally very small computers that cost next to nothing. They're found in toys like Furby, digital pets, talking dolls, and so on. It is not easy to deal with the suppliers of these computers; they demand huge orders.

  A look under the covers of a small robot made with such a chip is provided at www.phobe.com/furby/. Furby and Furbies are the intellectual property of Tiger Electronics.

Parallel Processors

Parallel processing is not new. The method stems from the realization that many computational problems do not have to be executed one step at a time. Often, a computational problem can be broken down into problems that can be executed simultaneously without fear that the work done on one problem will obviate the need for work on the other problem. In WWII, the atomic bomb project employed dozens of people who sat at mechanical calculators performing computations in parallel.

Most modern general-purpose processors (like those from Intel or Motorola) already contain more than one computer within the chip. This is done because almost every computational problem can benefit at least in some ways from parallel processing.

Consider for a moment the work done in the following software pseudo-statement: If A, then B, else C. The serial way to process this statement is to compute A, and then compute either B or C. With three processors at our command, we could compute A, B, and C all at the same time. When this single phase of computation is complete, the computer merely chooses (based on A), either B or C as the answer. This can save one computer cycle. It's true that a third of the work is wasted, but the program runs twice as fast.

The technique in the previous example cannot be extrapolated to a much more complex computational problem. As the statements in complex programs grow in size, the number of "branches" increases rapidly. In the previous If statement, only 1 branch was used, so we only needed about 2^1 processors (3 actually). In more complex programs with many branches, the number of processors needed grows very rapidly, making parallel processing impractical. The way to avoid this problem is to restrict the number of applications we try to tackle with parallel processing.

Many classical computational problems can still be partitioned naturally into parallel tasks. Consider weather processing or vision systems (for the robot). The field of view can be partitioned into areas, and a single processor can be assigned to each area in an array. Each processes the information coming into its area. Generally, the processors can communicate with their neighboring processors. In a weather application, each processor updates the weather in its small area (which may only be a few hundred meters square). It communicates with its neighboring computers to inform them about relevant events, such as moist air moving into their area. In such a way, weather forecasts have been made much more accurate and timely. The array processor has the general structure shown in Figure 3-5.

Such an array can be built using general-purpose processors, but companies have created processors specifically designed for parallel processing. They contain communication structures and special instructions that make parallel processing more efficient.

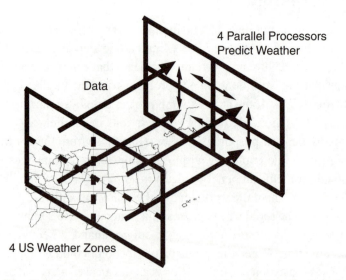

FIGURE 3-5 Parallel processors can divide up calculations for weather prediction.

Often, these companies support operating system software and compilers that make partitioning and hosting an application much simpler.

Here are a couple of URLs for further study on parallel processing:

- www-unix.mcs.anl.gov/dbpp/text/book.html
- www.afm.sbu.ac.uk/transputer/

Digital Signal Processing (DSP)

DSP chips are basically special-purpose processors designed to serve a particular class of computational problems. The central feature common to most DSP chips is a MAC, which stands for *Multiply and Accumulate*. And no, sorry, this has nothing to do with having lots of kids and living in a small house!

DSP processors are specifically designed to rapidly multiply two numbers together and add them to a third (accumulate). Several types of arithmetic problems are well served by such a processor:

- **Taylor series**   In 1712, mathematician Brook Taylor (see Figure 3-6) wrote a formula that can be used to approximate a function. Where f(x) is a function (with certain continuity restrictions) and $f^n(x)$ is the nth derivative of f(x) with respect to x, then f(x) can be approximated in the vicinity of x = a by the formula

$$f(x) = f(a) + f^1(a) \times (x - a) + f^2(a) \times (x - a)^2/2! + \ldots + f^n(a) \times (x - a)^n/n!$$

FIGURE 3-6   Brook Taylor

plus a remainder. This formula provides a standard way to approximate and compute functions like sine and cosine. It's the way compilers set up the computation. It involves several multiply and accumulate steps. Each term in the equation is another MAC. Generally, the remainder can be made arbitrarily small by carrying out more terms (making n larger). A tutorial on the Taylor series can be found at www.wikipedia.com/wiki/Taylors_theorem.

■ *Finite Impulse Response* **(FIR) filters**   These are generally used for filtering a continuous stream of information that represents audio or video. Consider the reception of an audio signal in the presence of a strong 1 kHz interfering noise source. We would like to remove the 1 kHz noise from our signal (as best we can). If the audio signal is digitized, it can be fed into a FIR filter specifically designed to filter out 1 kHz signals. The FIR filter method gives us a way to do this in as precise a manner as required, governed only by cost.

Suppose we want to filter the signal x(t) to produce signal y(t). The generalized formula for an n-stage FIR filter is given by

$$y(t) = h0 \times x(t) + h1 \times x(t-1) + h2 \times x(t-2) + \ldots + hn \times x(t-n)$$

where h1 . . . hn are the coefficients of the filter. We'll explain the math in a later chapter, but we can see that this formula is also a series of MACs. A web site on FIR filters can be found at www.wwc.edu/~frohro/qex/sidebar.html.

■ **Fourier Transforms**   Fourier Transforms were developed, as we might guess, by Joseph Fourier (see Figure 3-7) in the early 1800s. The transforms are a way of representing any function, within certain bounds, as the superposition of a series of pure sine waves. In this way, a function is broken down into a series of pure fre-

FIGURE 3-7   **Joseph Fourier**

quencies (multiplied by coefficients). A good drawing of superimposed sine waves can be found at www.yorvic.york.ac.uk/~cowtan/fourier/ftheory.html.

The Fourier Transform has many variants, including the *Fast Fourier Transform* (FFT) and the *Discrete Cosine Transform* (DCT). These transforms are commonly used to remove noise and unwanted frequencies from an image or signal as follows. The image is transformed into a series of discrete frequencies. Then the unwanted frequencies are erased (or the wanted frequencies are picked out). Either way, the wheat is separated from the chaff. Then the inverse Fourier Transform is computed to reconstruct the image, which is clearer and easier to understand than the original. Suffice it to say, the FFT, and other transforms like it, use a series of MAC operations.

In robots, FFTs can be used to identify objects in the field of vision. If FFTs are performed on the digitized field of view, the robot's DSP computer can look for the FFT signatures of specific objects, rejecting all those objects that don't conform. Interesting information on Fourier transforms can be found at www. yorvic.york.ac.uk/~cowtan/fourier/fourier.html and at www.medialab.it/fourier/ fourier.htm.

Notes on DSP

DSP processors have special purpose hardware that speeds up the computations they must perform. These hardware structures povide both increased accuracy and faster execution.

Arithmetic   We've seen that one of the central features of a DSP processor is the MAC, a hardware structure capable of executing a multiplication followed by an addition. This arithmetic operation is performed on a digital representation of a number. Numbers can be represented within a computer in a fixed-point format or a floating-point format. Be aware that DSP processors come in these two versions and that the floating-point DSP processor is much more expensive.

Fixed-point numbers are familiar to us as integers. A 16-bit fixed-point number can represent a of $2^{16} = 65536$ numbers. This range covers about 5 decades of range ($< 100,000$). But there are some problems with fixed-point format. If we were to multiply two fixed-point numbers like $60,000 \times 50,000$, the answer could not be represented in 16-fit fixed-point format.

To solve such an overflow problem, we temporarily can invent a "16-bit floating point" format. Such a format is impractical, but illustrative here. Many people are familiar with scientific notation where a number can be represented as 2.71×10^{12}, a very large number. Suppose we take our 16-bit number and divide up the bits differently, using 10 bits as the "mantessa" to represent the 2.71 number and 6 of the bits as the "exponent" to represent the 12 numbers in our example. This gives our floating-point numbers a range of about $2^{10} \times 10^6$, much larger than 65536. However, the accuracy is only $2^{10} = 1024$ instead of 65536. Our multiplication example from above ($60,000 \times 50,000$) can now be done because it does not overflow $6 \times 10^4 \times 5 \times 10^4 = 30 \times 10^8 = 3 \times 10^9$.

The floating-point formats used in computers are a little different than this. Please visit the URLs for a better description.

Floating-point gives us a wider range of numbers over which the arithmetic can take place. The differences between these two number formats are explained at web sites www.research.microsoft.com/~hollasch/cgindex/coding/ieeefloat.html and http://ee. tamu.edu/matlab-help/toolbox/fixpoint/c3_bev2.html.

DSP Hardware Many of the arithmetic problem domains we've looked at involve many MACs. The Taylor series, FIR filters, and FFTs all require the repeated multiplication of coefficients by data values to form a long summed-up equation. DSP processors have memory-addressing structures and control hardware that significantly speed up such repetitive operations. In math parlance, they are well suited for vector and matrix arithmetic. The most sophisticated also employs parallel processing to speed up these calculations.

DSP processors are often used to process continuous streams of information such as audio, video, or data from an RF receiver. The data stream never stops and must be processed at all times. Accordingly, DSP processors can have buffering built into their processing streams and avoid traffic jam interruptions that can stall a general-purpose *central processing unit* (CPU). Think for a moment about a desktop computer. How often does it lock up while performing some housekeeping task? Such lockups are not allowed in the processing of continuous stream data, and DSP processors can make sure that does not happen. If the robot needs to process continuous streams of media-type data, consider a DSP processor as an alternative.

Here is a PDF file and a few web sites on DSP processors and what they can do:

- http://bwrc.eecs.berkeley.edu/Publications/2000/Theses/Evaluate_guide_process_archit_select/Dissertation.Ghazal.pdf
- www.bores.com/courses/intro/chips/index.htm
- www.wave-report.com/tutorials/DSP.htm
- www.jovian.com/tutorial/demos.html

General-Purpose Processors

The primary advantage that general-purpose processors have is their speed. They can perform simple operations with blinding speed, and so complete great amounts of work. How do we go about finding the right one for our robot?

Computers came into being during World War II. They were made using vacuum tubes and were built in an effort to break enemy codes. Here's a nice site covering the history of computers: www.eingang.org/Lecture/index.html.

Not surprisingly, the best choice for the robot is the cheapest computer that gets the job done. Many design variations exist among the hundreds of computers that are available. To choose the best computer for the robot, we need to be well acquainted with the innards of the machines. This will give us a better perspective when the time comes to choose.

Computers have basic characteristics and architectures that have been worked out over the years. We'll take a look at each in turn.

WORD SIZE

Computers have, within them, the equivalent of a natural word size. They store and manipulate digital data that is represented by n bits, each representing a 1 or 0. An 8-bit computer has 8-bit words that store numbers from 0-255. A 16-bit computer has words that store numbers from 0-65535. The word size of a computer tells you the innate capability of the computer to manipulate numbers easily. The larger the word size, the faster the computer will be able to handle calculations involving large numbers. The first modern computer chips were 4-bit machines. I guess marketing didn't like the sound of selling 2-bit computers! All the internal structure of the 4-bit computers (the details of which we'll get to later in the chapter) were 4 bits wide, just enough to store the numbers from 0 to 15 decimal. That's great for counting the moons of Neptune (8 moons), but not Jupiter (47 moons and counting). To count Jupiter's moons, a 4-bit computer would need to use 2 of its words (8 bits), which would give it a capacity to count 256 moons. A 4-bit computer can still do the work, but it will be slower than an 8-bit computer at the same job because it has to do at least twice as many operations.

Modern microprocessors that we could use in our robot range between 8- and 64-bit word sizes. The 8-bit computers are generally well suited for most simple robot calculations and control system loops, but it's not a very expensive proposition to look at 16- and 32-bit computers. Computers with 64-bit word lengths begin to get pricey. One must look at a few central considerations when choosing the word length of the computer for the robot. Most robot designs have 8-bit processors to save power and money.

- **Data length**   How well does the word length of the computer match the data streams that the robot will have to deal with? If the computer is gathering vision data in 16- or 24-bit words, consider using a 32-bit computer. It is not unlikely that we'll have to perform 32-bit arithmetic anyway. If all the data gathering inside the robot generates 8-bit data, consider an 8-bit word length. But look closely at the arithmetic required. Be aware that even a simple addition of data can engender the requirement for extra bits of word length. If we add two 8-bit numbers together, we may well need a 9-bit number to store the result! Stepping up to the next largest word length computer is often a safe bet; a 16-bit computer might be needed.

- **Computer horsepower**   Even a tiny 4-bit computer can perform all the calculations required in a robot control system. The real question is, can such a 4-bit computer do it fast enough to keep up with the requirements of the robot? If we design the robot very carefully, we can minimize the requirement for a lot of computer horsepower. We can go into how to do that in a later chapter of this book. The point is, if we're sizing the computer to the task at hand, we can gain a lot by minimizing the task. Then we only have to pick a computer large enough to do the job.
- **Memory size**   Often, the word width of the computer dictates the word width of the memory bank. A 32-bit computer works best with a 32-bit-wide memory module. As such, the word length can also affect the size and cost of the memory.

POWER

Many robots are battery powered. We'll tackle power considerations later but should mention it here. To save power, look for the following features in a computer:

- Lower-voltage electronics
- Low-power operation
- Support in the operating software for low-power states
- Lower-frequency operation (if we can stand the slower operation)

MEMORY SUPPORT CIRCUITRY

Computers require memory to store their programs and data. The memory can be attached to the computer in several different ways. This section outlines some of those options.

- **Stored program**   Many questions have been asked about the program software itself. Where will it be stored? Flash memory and disk are two popular methods. Flash memory is more reliable physically, which is important if the robot will be mobile. We'll look at both types of memory shortly.

    Also, how will the program be changed? It's always a good idea to maintain the ability to upgrade the software in the robot. That means we need a method of getting the program information into the robot.

    This can be done in a number of ways, including through a communication channel. If the robot has a communication channel to the outside world, we can encode commands into the channel that will enable the reprogramming of the robot's software. If the robot is at a remote location (like Mars), we would have to do this very carefully. The accepted technique is to trigger the download command, pull in blocks of program data with full error detection and correction, store the program away in block form until it has all arrived, and then blast it into flash memory or disk. If possible, put paged flash memory in the robot so a boot program will always exist and will not change. The boot program can download and burn program flash. That way, we have a minimal chance of corrupting the program to the extent that we have no way to recover.

Another thing to remember about downloading over a long distance is that often significant communication delays occur. The downloading protocol has to survive all sorts of communication flaws, including long delays in transmission time. In the case of one of the Mars landing missions, the mobile robot could only be reprogrammed about once a day. In addition to communication delays, the reprogramming team had to put up with decreased communication bandwidth, planet rotation, sunspots, and so on. In general, make the communications protocols for the robot bulletproof. Expect the unexpected. Martians might even show up and stand in front of the antennae!

Sneaker Net is another way of getting the program information into the robot. If the robot is accessible, engineers can walk up to it and make the new software changes.

- **Memory addressing range** Computers have instruction sets that encode addresses; the instructions are stored in memory as a series of bits. This allows an instruction to directly access a memory location for reading, writing, or modification. To encode a memory address into an instruction, the address must take up some bits within the instruction. Often, some of the bits in the instruction will reference another register with many more bits to fill out the address. The final, resolved address is called the *effective address*. The number of different memory addresses that can be accessed at any one time depends on the number of bits in the effective address.

Different instructions of the computer will be able to access different ranges of addresses. By and large, the word length of the computer sets the largest address range. A 32-bit processor generally can address 2^{32} bytes (about 4 billion bytes). Processors with 8 and 16 bits generally use a 16-bit address range for 65K bytes. The memory addressing range is important because it restricts the number of memory bytes that the computer can see at any one time. If our robot's software is looking at many thousands of bytes at any one time, consider whether a 16-bit addressing range is sufficient. It does not cost a vast amount of extra money to step up to a 32-bit computer. If the computer has a *memory management unit* (MMU), it is possible to step up to a very large addressing range and to support a vast memory.

■ **MMU**   An MMU is a set of registers within the computer chip that enables the computer to access a vast memory array. Let's use a visual image to describe what an MMU does. Think of the memory array as a vast outdoor wheat field of bytes. Think of the computer as being inside a house with a window looking out on the field of bytes. The computer can process instructions to manipulate all the bytes it can see out of the window, but not the ones it cannot see. Now let's make a magical MMU that can move the window around the wall of the house. The MMU stores window locations and can remember a bunch of different locations for the window (called *pages*). In fact, each user of the computer can have his or her own window location and, as such, a private memory space out in the field of bytes. In this way, the computer can support multiple users without the difficulty of keeping them all apart. If only the operating system can manipulate the MMU, then it's possible to keep the users secure from one another so they cannot disturb each other's field of bytes. In a robot design, this can come in handy if multiple groups of engineers reprogram the robot's functions. It is possible to keep them from interfering with one another.

   ■ In addition, if a user needs more memory than the addressing range allows, a secure portion of the MMU can be made available to the user. The user can control multiple pages of memory to get access to more memory. The only catch is that the pages cannot all be accessed at the same time without altering the MMU between accesses.

   ■ So how does an MMU work? Basically, the computer must come up with extra memory bits to add to the largest address range, which can be done in several ways. In the first place, a few extra bits can be added by allowing multiple users to access the overall memory. Accommodating 32 users would add 5 more bits. Most computer architectures enable each user to control a few more bits. The net result is that the MMU structure, inside the CPU, looks just like a small memory. The address signals of the MMU memory is made up of the extra bits. The data stored in the memory is generally the effective address of the user's memory page. In addition, the MMU memory contains security bits that specify what sort of operations are allowed on the memory page. It is possible to disable writes and reads, and to restrict access to different classes of users.

   To recap, an MMU enables the computer to access a much larger memory than the addressing range ordinarily does. In addition, an MMU can provide security for multiple users. In general, unless the robot design is very complex with a large operating system and many users, an MMU won't be of much use.

MEMORY CHIPS

Oh yes! Most computer memories actually contain memory chips. These are integrated circuits that contain thousands or millions of individual bits that the computer can read and write. A few different types of memory are available, and they all bring different benefits to a robot project. It makes sense to know about the most popular types of memory and what they can do for the robot project.

Flash Memory

Every computer needs a place to store its operating program. The program must not vanish when the power goes off. With current technology, almost every computer contains some flash memory, which contains the initial software that the computer runs when it boots up. The same flash memory can contain the bulk, or all, of the computer's software program. Flash memory's primary advantage is that it retains its contents in the absence of power, making it nonvolatile memory. We won't go into the physics of it here.

Flash can be programmed when the robot is built and will retain the program throughout the life of the robot. Most flash memory can be reprogrammed in the field if the program must be changed. Beyond just storing the program of the computer, the flash memory can be used to permanently store other data the robot may gather, almost like a disk system.

One caveat, however, is that many types of flash memory can only be written to a specific number of times before failing. The flash memory chip specifications will detail how many times the flash can be written to. So if a need exists for nonvolatile memory storage now and then, consider putting flash memory into the robot. Sometimes this sort of memory can be added to a robot's computer using *Personal Computer Memory Card International Association* (PCMCIA) cards, which we'll talk about in a bit.

Static Memory

This is a type of volatile memory, which is relatively simple to use from an electrical engineering perspective. It does not require complicated timing. However, static memories are generally smaller for equal dollars and have fallen out of favor. They generally use two to four transistors just to store one bit of memory, whereas the cheapest (*Dynamic Random Access Memory* [DRAM]) memories use just one transistor to store a bit. One thing static memories are good at is battery backup. Static memories can be made nonvolatile with the addition of a battery. They are often teamed up with lithium or other such batteries that have a long shelf life. Some types of static memories consume very little battery power when they are off and can retain critical data for long time periods.

Dynamic Memory

Most computer boards these days use flash memory for the nonvolatile boot program and dynamic memory for the bulk of the volatile memory space. It's not uncommon for the entire computer program to be stored in flash memory, transferred to dynamic memory, and executed from there. The reason is execution speeds out of dynamic memory are often faster. To understand why, we have to go into the physics this time.

DRAM behaves the way it does for one primary reason: It only uses one transistor to store a bit. It does this by taking advantage of some of the capacitance under the transistor. A capacitor is basically a place to store electrons. The number of electrons in the capacitor determines whether a binary one or zero exists in the bit. A data bit, in the form of voltage, can be moved to the transistor. Then the transistor can put the data into the capacitor just by turning on. If the data, represented by voltage, is a one, then electrons flood into the capacitor. If the data is a zero, the capacitor is drained of electrons. When the time comes to read the data bit, the transistor turns on and the number of electrons in the capacitor is inspected. If enough of them are present, the computer reads a one.

DRAM is very dense because it only needs one transistor per bit, thus saving space on the integrated circuit itself. However, some problems occur with this memory structure. For starters, the very act of reading the bit destroys it. This is called *destructive readout*. Immediately after reading the bit, the memory support circuitry within the computer must rewrite the data bit back into the capacitor.

Another problem happens as well. Once a bit is written into the capacitor beneath the transistor, it begins to deteriorate. The electrons in the capacitor begin to leak away one at a time. It only takes a few milliseconds before the integrity of the data bit can be called into question. Accordingly, many of the memory chips have circuitry within them to automatically read every bit and rewrite it every few milliseconds. This process is called *refresh*. Some computers perform this operation using refresh circuitry within the computer chip itself. Be very careful to think through the refresh scheme when choosing memory for the robot. At least one of the chips must handle the refresh task.

One of the other disadvantages of DRAM is the complex timing required for the signals. We'll get into how DRAM works in a minute, but the complex timing of the signals brings up two problems. First of all, almost no way is available for putting the computer to sleep to conserve power. With all the signals running all the time, the DRAM generally cannot go to a low-power mode. If a low-power sleep mode is important for the robot design, consider SRAMS instead. Second, if we're building our own computer from scratch, be very careful to analyze the timing of the DRAM signals. If they are even off a little from the requirements, errors can occur that will be hard to isolate.

To use DRAM properly, we have to look into its internal construction. DRAM is commonly built as an array of bits. If a million bits ($1{,}024 \times 1{,}024 = 1$ million) are inside the DRAM, the bits may well be arranged as 1 large array with 1,024 columns, each of which has 1,024 bits in a row. The address lines coming into the DRAM generally are timeshared. To address 1 million bits inside the DRAM, 20 address bits are required ($2^{20} = 1$ million). Instead of having 20 address pins on the DRAM, it likely only has 10, and they are used twice in the following manner.

The first 10 bits of the address are presented to the DRAM. These 10 address bits can address an entire row of bits within the memory array. This cycle is called RAS for *Row Address Select*. During this time period, the entire addressed row of 1,024 memory bits is read into a RAS read register inside the DRAM. Next, the computer chip provides the remaining 10 address bits at the address input pins of the DRAM during what's called the CAS cycle for *Column Address Select*. During the CAS cycle, only one of the 1,024 memory bits from the RAS read register is sent to the DRAM output pin. This is the RAS/CAS cycle. This type of architecture saves a great deal of space and circuitry inside the DRAM and has become a standard in the computer industry.

The timing of all the DRAM signals must be very precise to avoid errors. Most computer chips on the market will drive DRAM directly with default timing known to work with contemporary DRAM. Most computer chips also have registers within them that can be used to change the default timing on the computer chip's DRAM interface pins.

One of the interesting benefits of the RAS/CAS cycle is that, in our example, 1,024 bits are fetched at the same time during the RAS cycle. It's only a preference that we happen to want only one bit during the CAS cycle. The truth is, if we run multiple CAS cycles after the single RAS cycle, we can fetch many bits out of the RAS read register. This method of using DRAM is generally called page mode, and not all DRAM supports it. The next section dealing with cache memory will illustrate a good use for this feature.

DRAM comes in many different styles, each with a different acronym. They each have different timing and power requirements. For further study, check out www. arstechnica.com/paedia/r/ram_guide/ram_guide.part1-1.html and www.howstuffworks. com/ram.htm.

CACHE MEMORY

Great, just when we thought we had this memory thing licked, along comes another kind. Cache memory (pronounced "cash") is a small amount of memory within the computer chip that greatly speeds up the execution of a program. The central idea is that the DRAM memory chips external to the computer chip take a good long time to deliver their contents to the inside of the computer chip, maybe 60 ns. That may not seem like a long time, but if we consider that the computer chip may be able to execute instructions every 10 ns, it does waste a lot of time waiting for instructions to come out of memory.

What the cache does is watch the access to external memory. If the cache control circuitry inside the computer chip believes it already knows what the contents of the memory address are, it cuts short the computer chip's memory cycle and simply pulls the data out of its own cache memory instead. This way, the instruction will be executed two to six times faster. It's easy to use cache since it's transparent to the programmer. The cache is simply turned on, and it automatically functions to speed up the program execution.

Many computer programs will execute in tight loops for short periods of time. The execution of a FOR loop in C is a typical example. FOR loops will execute the same instructions for a prescribed number of iterations. While executing in a FOR loop, a C program will execute the same instructions over and over again. If these instructions are put into the cache memory, the FOR loop will execute much more rapidly. As a general rule, most programs will execute in such "local" loops a large percentage of the time. This is the true power of using a cache memory structure within a processor. Even a small amount of cache memory goes a long way. Generally, only the faster computer chips have cache circuitry since only they can truly take advantage of it.

How does cache memory work? First, we'll describe a more complex structure for cache memory; later we'll look at a simplification. First of all, cache memory usually has just a few thousand words. Each of these words can contain both a full memory data word (duplicating the contents of a DRAM memory address) and the DRAM memory address itself. As the computer reads data from a DRAM address the first time, the cache memory controller puts the data and the address into the cache memory at the same time. Later, if the computer program reads that DRAM address, the cache memory recognizes the address as a match, gets the computer's attention, rapidly substitutes the data from the cache, and cuts the memory access short. As the program continues to access DRAM addresses in a small "local loop," all the data from those addresses is also put into the cache memory. As the program continues to loop through those DRAM addresses, the cache memory steps forward with the data and acts to speed up the computer. When the program moves on to another portion of the program, new data is cached.

But what happens when the cache fills up? Generally, the cache controller has hardware that examines the least used cache words. When a new location is required for cache data, the controller then selects the least used cache location, dumps the old, unused data from it, and puts the new cache data in it.

As a side note, when data is written into memory that is also cached, the data is written into the cache memory at the same time as it's written into the real DRAM. That way, the cache data remains the same as the contents of the DRAM. An article at www.pcguide.com/ref/mbsys/cache/func_Write.htm describes writing to cached memory locations. Follow all the links for a complete explanation.

The cache controller must recognize when it must act while the computer is accessing a DRAM address. The most complex method is to store the DRAM address inside the cache memory. The cache controller must then have address-matching hardware that can compare the computer-generated DRAM address with *all* the addresses within the cache memory bank. This type of hardware is expensive and is generally known as *Content Addressable Memory* (CAM). A less expensive alternative is simply to cache only within a small address range. If the computer can cache all the DRAM data that resides within a certain memory address range, things are simplified. The cache controller need only compare the upper bits of the computer-generated DRAM address with the address of the cached memory range. The cache memory controller and how it recognizes situations where it comes into play are discussed at www.pcguide.com/ref/mbsys/cache/func_Mapping.htm.

Cache memory can reside in a few different places. If it's inside the processor chip, it's generally termed a *Level One* (L1) cache. It's the fastest and, because it's inside the computer chip, it's generally the smallest and most expensive.

Board designers can also put cache memory chips between the computer chip and the DRAM. This cache, external to the processor chip, is generally called a *Level Two* (L2) cache. Sometimes the L2 cache is also inside the computer chip. The L2 cache is slower than L1, but it is often bigger. The cache controller looks to the L1 cache first. If the L1 cache does not have the data, the cache controller looks to the L2 cache. If the L2 cache does not have the data, the cache controller goes to the DRAM.

The following web sites help define L1 and L2 caches. Follow all the links; there is much more to learn:

- www.pcguide.com/ref/mbsys/cache/
- www.pcguide.com/ref/mbsys/cache/role.htm
- www.pcguide.com/ref/mbsys/cache/layers.htm
- www.pcguide.com/ref/cpu/arch/int/comp_Cache.htm
- www.computerhope.com/jargon/l/l1.htm
- www.computerhope.com/jargon/l/l2.htm

Cache Thrashing

As we've seen, cache memory is most effective when the computer program loops in a small local loop, a portion of the program confined to a small number of DRAM memory addresses that can all reside in a cache at the same time. It is possible to misuse cache memory. Consider a program that skips around all over the place in memory. The cache controller cannot be effective if it cannot store all the instructions in the cache at the same time. It is continually asked to put new locations into the cache and is ineffective. The programmer is said to be "thrashing cache." Say that five times fast!

Be careful in the design of the robot's software that the program execution does not jump around too much. In larger, more complex computer chips (such as StrongARM), it is possible to confine the use of cache memory to specific memory ranges and thus avoid areas of the computer program that will dump the cache without positive benefit.

By the way, cache memory can also hold and mimic the contents of flash memory too. This is useful if the processor executes out of flash.

Cache Interaction with DRAM

We mentioned before that DRAM can be used in paging mode. When a processor with a modern cache controller (like the Xscale StrongARM) reads a DRAM address, it does not simply read just one instruction. Since the DRAM retrieves 1,024 (or so) bits at a time during the RAS cycle, the processor can execute, for example, 16 CAS cycles to fill up the cache with the subsequent words from memory. This is a very time-efficient way to fill the cache memory. The processor is up to other things with the fetched instruction while the cache controller is busy dragging words out of memory with page-mode CAS cycles. The only drawback of such activity is that it makes it difficult to monitor the actions of the processor externally by just observing the activity on the memory address lines.

COMPUTATION AND STORAGE REGISTERS

Every computer chip is capable of performing arithmetic and logical functions. They contain computational circuitry that can add and subtract word-length words at instruction speeds. Certainly, it's important to analyze the requirements for the robot and the arithmetic computations that will be necessary. We can talk about that in another section of the book, but it's important to note one or two things here.

First, computers contain spare, word-length registers that are used to store intermediate results when they are not in use. If a computation handles many different numbers at the same time, a computer with many spare registers (termed *general-purpose* [GP] registers) can often execute the computations at a faster rate. To take advantage of this capability, we often have to take a very close look at the software and the compiler (if one exists). Often, a compiler will automatically avoid using GP registers, preferring to use slower memory locations instead. This is done so the compiler will be usable on many different computers, some of which have few GP registers. If we have specific software routines (like loops in the robot's control system software) that we want to speed up, we can pay specific attention to that small area of code. Often, with C language statements, such as the *register* construct, we can force the compiler to generate code that will use the faster GP registers during computation. We still have to examine the intermediate assembly code to make sure we are getting the results we desire. Certainly, if the robot's code is written in assembly code, we can force the issue much more easily. The point is, consider the internal register structure of the computer when picking the computer or designing the software.

Second, be advised that some computers have more computational hardware than others. All computers have fixed-point computational capabilities, and some have floating-point capabilities as well. Others, as we have discussed, have very special-purpose compute units with DSP or communication hardware built in. Again, take a close look at the computer requirements.

INSTRUCTION SET

An instruction set is the base language of the computer. These are generally word-length, assembly language words that the computer can look at to understand what it must do during the execution of a program. It does not matter whether the program is written in C, Forth, C++, Fortran, or assembly language. The compilers and assemblers always must reduce the program to a series of instruction-set commands that the computer can execute. In assembly language, they look something like this:

| ADD | r0, r1 | (Add GP register 0 to GP register 1) |
| SUB | r2, r3 | (Subtract GP register 2 from GP register 3) |

When translated to binary, they would reside in instruction words like this (to use an imaginary 8-bit computer instruction set):

Bit		7	6	5	4	3	2	1	0
OpCode		x	x	x	x				
Source register						x	x		
Destination register								x	x
where the OpCodes are									
0000 for add									
0001 for subtract and so on									
ADD	r0, r1 codes as	0	0	0	0	0	0	0	1
SUB	r2, r3 codes as	0	0	0	1	1	0	1	1

These instructions are decoded at very high speed within the computer and are executed immediately.

Years ago, computer companies built computers the size of refrigerators and they tried to sell customers on the richness of their instruction set. The truth is, almost nobody cares about that. People buy computers based on almost every other reason other than this. So why should we care?

If we're building a robot and we're watching our budget, we should select the computer carefully. Some computer chips will match our requirements better than others. Looking at it the other way, given a more powerful computer chip, we can often make savings by tailoring the robot's algorithms to the power within the computer. So let's take a look at some of the wrinkles that have come along in instruction sets.

RISC

RISC stands for *Reduced Instruction Set Computer*. The imaginary 8-bit instruction set shown earlier is similar to a RISC instruction set. The instructions are elemental and can generally only perform one small computation at a time. RISC machines were supposed to get their power from blinding speed, even compared to computers with more complex instruction sets.

RISC computers were touted a decade ago as a major advance in computer hardware, designed to significantly speed up computations. The technology did not, by any means, take the computer world by storm. Many RISC computer designs are still around, such as MIPS, ARM, and others. They generally have smaller, simpler semiconductor dies and can be incorporated into ASICS much more easily than larger computer cores. But they have found not found their niche because of their speed. Rather, they've found their place in low-power designs and in the relative transportability of the designs. One of the best ways to see what an RISC computer can do for the robot design is to simply look at several other designs the computer chip has been used in.

Here are some web sites on the history of RISC computers, discussing them in greater depth. In addition, some articles show the advantages of RISC over *Complex Instruction Set Computers* (CISCs), which we will talk about shortly. Some of the articles are years old but still have a relevance:

- http://copland.udel.edu/~anita/risc.html
- www.cs.washington.edu/homes/lazowska/cra/risc.html
- www.appliedembeddeddesign.net/design_riscCisc.asp
- www.ccs.neu.edu/groups/honors-program/freshsem/19951996/utopia/risc.html

CISC

Complex Instruction Set Computers (CISC) have been the norm since the commercial introduction of computers. The instructions are much more complex than those in RISC machines (hence the name). Although a CISC machine may still have ADD and SUB (subtract) instructions, it may also have MPY (multiply), DVD (divide), ECC (error checking and correction), and MAC (multiply and accumulate) instructions that perform complex calculations. A MPY instruction typically requires a series of ADDs and SHIFTs. A DVD instruction requires a series of SUBs and SHIFTS. A MAC requires at least an MPY and an ADD.

It can be very expensive to build the control circuitry within a computer that can manage the cycles in such a complex instruction. What most processor designers did was build a uCode (microcode) engine into the computer processor. Effectively, a small, very high-speed uCoded RISC processor would be inside the CISC processor. The uCode program would reside in high-speed *Read-Only Memory* (ROM) and would execute a very short series of uCoded machine cycles to carry out the intended CISC instruction. The uCode for an MPY x,y instruction would look very much like the following:

```
        uLOAD          z, wordlength     ; MPY instruction execution.
Loopr:  uSHIFTRight     x                ; Shift x
        uTSTskp Carry   x                ; Was a 1 there?
        uADD            x,y              ; Yes, add Y to answer
        uDECJmp         z, Loopr         ; No, loop til done
        uRTN . . .                       ; Finished
```

The uCode program would execute a series of shifts and adds to accomplish the MPY. The advantage of uCoding the instruction set is that the CISC hardware could be simplified. Many CISC instructions could be coded with just a few entries in the uCode ROM. A CISC instruction set might take longer to execute a program, but the compiled C programs (supplied by the users) would have a smaller number of bytes.

Some uCoded processors enable end users to supply uCode, which can be executed out of fast RAM inside the uCode engine. This feature is an attempt by the CISC designers to capture some of the advantages of the RISC architecture. End users can effectively make their own instructions up. This is of use if the robot has one or two simple algorithms that must run faster. It can be very difficult, however, to write uCode. The documentation is often not very good, and support is often worse.

The major advantage of a uCoded CISC machine is in the richness of the instruction set. Many of the CISC machine designers provide specialized instructions that can be of great use in specific circumstances. Some CISC computers will have specific instructions for the treatment of continuous streams of data such as might come from communication interfaces. This becomes almost a crossover capability from DSP machines. If the robot needs specialized communication or data-processing instructions, look for them in the CISC instruction set of the processors under consideration. Pentium™ processors have the *Multimedia Extensions* (MMX) instruction sets and certain constructs that are good for the processing of vector data.

Some CISC machines provide floating-point instructions, which can greatly speed up some algorithms. Others provide communication instructions such as the computation of ECC polynomials and Viterbi codes.

COPROCESSORS

In an effort to provide extra horsepower for their processors, some designers couple processors together. The Pentium™ class processors can operate in tandem. This allows the same programs to operate a bit faster without significant modifications. For more information, go to www.intel.com/products/server/processors/server/xeon_mp/index. htm?iid=search+XeonMP.

Other designers coupled together processors with disparate capabilities. The PowerPC™ from Motorola is in such a class. The second processor is termed the communications processor and is reserved almost exclusively for the use of Motorola processor designers to provide communication processing. It's a simple RISC machine that is not documented for end users. It's used, for instance, to provide the processing necessary to implement LAN (local area network) communication interfaces. The communication processor can handle several communication interfaces at the same time, limited only by the overall bandwidth of the coprocessor. Other communication protocols like ATM, Sonet, and others are available as complete uCode that can be loaded into the coprocessor and kicked off.

If we can find a processor with significant coprocessing power, it can be used for parallel processing. The newly introduced network processors can be used like this. They are basically multiple RISC processors in a single chip. They're used for the processing of packets on the Internet, a task that can be partitioned and requires great horsepower. These chips are available from Vitesse, IBM, Motorola, and many others. Thus far, they have been used only for network processing, which handles IP packets in real time, but they are very powerful parallel processing machines and might work well in a robot control system.

INPUT/OUTPUT (I/O)

No matter how good a processor is, it's useless if it cannot communicate with the outside world. A computer can only process information as fast as the slowest link in its communication chain. Besides traditional *input/output* (I/O), which we'll get to in a moment, other communication paths within a computer can slow it down. If the memory is too slow for the processor, everything slows down. When designing a robot system the first time, be very careful to analyze the required communication bandwidths throughout the computer circuitry.

Think of a person who is blind. A blind person can certainly think fast and figure things out, but might take longer than most to assimilate visual or printed matter through Braille. As such, a blind person might not be the best choice to be an air traffic controller. I know I'm going to get into trouble for that statement, but it's true. The world is put together to suit sighted people and thus often puts blind people at a disadvantage. However, blind people might have advantages in situations where their specially trained hearing skills come to the fore.

So, too, certain processors have more I/O bandwidth than others. If the robot's system architecture calls for a processor to digest and process all the bytes coming in a 1000BT LAN interface at full speed, it's a pretty good bet there won't be any 8-bit processors that can handle it.

We must evaluate many places inside the computer hardware to determine if enough bandwidth will be able to handle the contemplated design. The list of considerations includes the following computer components.

Buses

A bus is a communication path within the computer that carries data from one place to another. Generally, a bus is a collection of wire traces on the PCB with a protocol that defines the meaning of the signals on the traces. It is not possible to put more data across the bus than the protocol claims it can handle. In fact, it is rare that the full, raw bandwidth capability of the bus can ever be achieved. While planning the design of the system, it is wise to derate the bus to 50 to 80 percent of the advertised bandwidth. This is likely the fastest speed at which we will be able to drive bytes across the bus.

Buses are generally designed by industry committees to solve particular data transfer problems. Often, the bus will have been designed to enable multiple manufacturers to build compatible equipment. Buses have characteristics such as width (analogous to word length), bandwidth in bytes per second, voltages, and loading. Loading defines the number of separate devices that can be connected to the bus at the same time.

Multiple buses are used inside most computer systems. To make sure we compare the bandwidth across each bus to the requirements set by the system architecture, it's important to list every bus within the computer. A few of the buses are hidden from most users and do not even have popular buzzword names. Let's look at those buses first, and then proceed to the popular, named buses.

Memory Bus

All computers have an interface between the memory and the processor. The processor can only read and write to the memory at specific speeds. If the system architecture calls for the processor to read data, manipulate it, and rewrite it, then we must be very careful about the memory bus speeds. If the system architecture calls for the data manipulation to be completed within a strict time budget, we must add the processor execution time to twice the data transfer time (one transfer to read, and one transfer to write). The transfer times may turn out to be a significant percentage of the overall timing. If this is a problem, we can look for a bigger memory bus to work with.

Sometimes the processor is just too slow on the transfers. If that is the case, we can look for *Direct Memory Access* (DMA) hardware. DMA circuitry can transfer bursts of data faster than most processors can. Sometimes DMA hardware is included within the processor, and sometimes we can add it on with external chips. Smaller processors will generally not have DMA capabilities. Here's a good rule of thumb. If the analysis of the robot's architecture shows that the memory bus is loaded down by as much as 30 percent from data moving across it, consider a faster computer, a wider memory bus, or DMA transfers.

Video Bus

Many computer systems are used to process vast amounts of video or graphics data. Game systems certainly are like this, and specific computer graphic buses are very fast and flexible. The penalty for choosing the wrong graphics bus would be poor graphics, delayed images, or system failure. If the robot design will use a great deal of graphic display and manipulations, consult the following site as a start: www.agpforum.org/.

Many other named buses exist within a computer as well. The following URLs are part of a superior web site. It's a great place to start, comparing buses and looking up their specifications:

- www.interfacebus.com/Bus_Design_Top.html
- www.interfacebus.com/Design_Interface_table.html

Here's one computer bus almost everybody overlooks: www.hits.org/hits/bus/bus5.html. Hopefully, you'll see it coming!

Now let's talk about some standard buses. More information on each can be found in the two previous interface.com web sites. The standard buses are as follows:

- ***Industry Standard Architecture* (ISA)**   ISA no longer means what it says. This bus came out with the original PC and was the mainstay of the industry for many years, but it's obsolete in that industry now. The bus had a limited bandwidth at 8 MBps. Don't use it! For more info, go to www.interfacebus.com/Design_Connector_PCAT.html.

- ***Peripheral Component Interface* (PCI)**   The PCI bus has taken over as the standard bus in the PC industry. It's a bus with a specialized type of signal that is limited in two ways. Signals can only traverse a limited distance (roughly the size of a PC motherboard). In addition, only about four loads (like connectors or integrated circuit pins) can be put on the bus before it starts to load down and fail. Bridge chips exist that can extend the PCI bus to more loads and sockets.

  A few versions of the bus exist, differentiated by the voltage, word width, and frequency. The most widespread version has the following characteristics: 5 volts and 32 bits at 33 MHz. This gives a bandwidth of $(32/8) \times 33$ million = 132 MBps per second (raw speed). As a practical matter, nobody could ever get better than about 100 Mbps out of the bus because of housekeeping tasks that take place on the bus. The maximum size of PCI bus technology lately is 64 bits at 133 MHz for a 1 Gbps bandwidth (raw speed).

  PCI has become an industry standard. Many board manufacturers and many chip manufacturers have adopted it. If the robot's computer supports the PCI bus, many third-party boards will be available to customize the design and save time (see www.interfacebus.com/Design_Connector_PCI.html).

  The PCI bus would be an excellent choice for a robot as long as the vibration problems can be addressed. The bus has around a hundred pins on each connector. It only takes one pin to fail from a vibration to bring a system down. If reliability is a key, look into the Compact PCI standard. It's a bit sturdier (see www.interfacebus.com/Design_Connector_CPCI.html).

- **PCMCIA cards**   This standard describes not so much a bus as an interface socket. Many peripherals are available as pocket-sized PCMCIA cards, so it's a good option for adding memory and peripherals to a robot. Most portable laptop PCs have PCMCIA sockets to accommodate these cards. The transfer rate is on the order of 20 MBps (see www.interfacebus.com/Design_Connector_PCMCIA.html).

- *Universal Serial Bus* **(USB)**   USB is a serial standard (using a thin cable) that is capable of transfers at around 1.5 MBps. It's well known in the PC industry and enables peripherals to be plugged in and out of the computer quickly, even with the power turned on. For robots, a USB might be an easy way to hook into another computer for communication or downloading. Many portable PCs support this standard and could be brought up to the robot to service it. For more info, go to www.quatech.com/Application_Objects/FAQs/comm-over-usb.htm.

- **Firewire, IEEE1394**   The Firewire standard is generally used in systems requiring a great deal of media data (audio or video). Cameras and other media devices connect together using thin, hot-patch serial cables. Audio and video can be transferred in real time, without interruptions, between devices. One peculiarity of media streams is that they cannot be interrupted without a noticeable degradation of the transmission. If an interruption occurs in a digital video stream, for example, blocks can be seen on the screen. The Firewire protocol is designed to guarantee the delivery of media data across the timeshared wire. If the robot must transfer video or audio data, Firewire might be a good candidate for those transmissions. Broadcast video requires a transmission bandwidth of around 15 to 34 MBps. Firewire can handle around 50 MBps (to accommodate multiple transmission streams) and faster versions are planned. For further info, go to www.interfacebus.com/Design_Connector_Firewire.html.

- *Controller area network* **(CAN)**   The CAN bus is a serial bus standard designed for use in electrically noisy environments such as automobiles and industrial sites. It can transfer data at up to 125 KBps over cables from 40 to 1,000 meters long (depending on data rates). Its other major advantage is that it saves wiring cost, an important consideration when making thousands of automobiles. If the robot generates a great deal of electrical noise from its motors, then CAN might be a good choice for the electrical bus inside the robot (www.interfacebus.com/Design_Connector_CAN.html).

- *Inter-IC* **(I2C)**   I2C is a serial bus standard used largely for signaling within a single PCB, although at least one version of the protocol uses a cable. Its primary value is to save pin count on small chips that require slow, complex data transmissions. If the robot is short on PCB real estate, then I2C chips can save quite a bit of room. The maximum bandwidth is around 400 KBps (www.interfacebus.com/Design_Connector_I2C.html).

Network

No discussion of I/O would be complete without a discussion of LAN I/O. Almost every computer system has a network interface, whether it's hooked up or not. In everyday business life, every computer is hooked up to the network in the office. The network, as it applies to a computer, looks just like a single cable that hooks into the back of the computer. The computer software knows how to talk to the other computers on the network and can use the wire to do so. We will discuss just how this occurs in the chapter on communications. The commonly used communication links (between computers) used in network communication are the following:

- **10/100BaseT** A single cable attaches to the back of the computer to provide 10/100BT connectivity. The cable may look like a phone cable or it may be coaxial like a cable TV cable. It may well use the Ethernet signaling voltages and protocols, and it probably is directly connected to a switchbox in a back room full of server computers. It's the single most popular way of connecting computers together and would be a good choice for the robot. Most computer boards that are purchased off-the-shelf have a network connector on board, but beware; this type of communication system requires a sizeable amount of software to support it. If the robot needs such a communication link (and connector), make sure the robot's computer will include the proper operating and network stack software. We'll discuss this further later.

  A 10BT interface has a raw bandwidth of 1.25 MBps but generally cannot support more than 75 percent of that. The 100BT is 10 times faster. Many computers support both interfaces. A 1000BT interface is 10 times faster again, supporting a bandwidth of 125 MBps. Don't forget to derate this number for practical purposes. However, at such speeds, many computers cannot even keep up with a 1000BT interface; high-speed, 32-bit systems are needed. Remember, the robot's computer system will only be as fast as its weakest link. Here are a couple of web sites about the type of wires needed for such communication links:

  - www.linksys.com/faqs/default.asp?fqid=18
  - www.zytrax.com/tech/layer_1/cables/tech_lan.htm

- **Wireless (RF)** It's not unlikely that the robot will need to be mobile. Assuming that's the case, having a *local area network* (LAN) wire plugged into the wall is impractical (and not cool). Designers long ago freed the owners of portable PCs from such wiring tethers with the introduction of a few wireless protocols. The most popular is 802.11, which comes in various versions, the most prevalent of which is 802.11b. It boasts speeds of up to 1.3 MBps, comparable to the wired 10BT standard.

  If we can restrict the communication needs of the robot to a fraction of that bandwidth, then 802.11b might make an excellent choice for external communications. Just be aware that an RF communication link is much more prone to errors than a wired link. Motors, computers, lights, radio stations, and even the stars all put out interference that can quickly corrupt a well-designed, standard RF communication link. Be very careful when using RF links. Make sure the transmission distances are well known because RF signals degrade rapidly over distance. The control system for the robot must be capable of surviving the interruption or corruption of the data streams coming over the communication link.

Here are a few sites about 802.11 technology:

- www.computer.org/students/looking/summer97/ieee802.htm
- www.wave-report.com/tutorials/ieee80211.htm
- www.homenethelp.com/802.11b/index.asp
- www.80211-planet.com/tutorials/article/0,4000,10724_1439551,00.html
- www.webopedia.com/TERM/8/802_11.html
- www.intelligraphics.com/articles/80211_article.html

- **Wireless Infrared (IR)** IR light is another possible method of communicating from place to place. IR exists in sunlight and everywhere in our daily existence. It can give us sunburn and it's just waiting to ruin the first robotic IR communication link it can find. Stray IR radiation is less prevalent indoors and has been used indoors for low-speed data links over short distances. But even indoors, RF communication links are beating the pants off IR communication links in all respects. The TV clicker may be the only appliance still using IR inside most houses.

Here's a great site detailing much of the *wireless LAN* technology currently available: ftp://ftp.netlab.ohio-state.edu/pub/jain/courses/cis788-95/wireless_lan/index.html.

PERIPHERALS

To round out our talk about I/O, let's talk about peripherals. Although the use of peripherals involves data communication over communication buses and links, it differs in some respects. Peripherals are often thought of as sources or destinations for data. They are devices that are attached to the computer to allow the entry, storage, or display of data. Peripherals are a bit boring and commonplace; they're described in thousands of articles. So why talk about them here?

First of all, robots are generally thought of as portable devices, clunking away their existence in dusty, poorly lit industrial cubicles to satisfy the peevish desires of their slavish masters. (Does this hit home? I hope I haven't ruffled any feathers out there!) Peripherals have to be carefully chosen if they are to match the requirements needed for a robot. If the robot moves or vibrates, many new requirements must be addressed, including vibration, shock, temperature, humidity, power reliability, and electrical radiation. We'll look at all these factors later as we consider environmental issues, but we can take a look at some of the I/O peripherals in this chapter.

Disks

Hard disk (HD) drives are familiar to most people. They are in most personal computers and are occasionally a source of frustration if they misbehave. When they break, they can cause significant headaches and losses of expensive data. The environment they operate in is important to consider. An office environment is quite stable in most respects. If an HD is put into a robot, it must be treated properly. Readers should extrapolate the following discussion to CD-ROMS and other rotating media since the following discussion will only address magnetic HD disks.

An HD is basically a spinning disk of magnetic material that can contain bits on its surface. A read/write head glides over the surface and provides access to these bits for the computer. When designing an HD into a robot, consider the following HD characteristics:

- **Vibration**   The HD, while it's running, holds its read/write head over the disk surface. The distance between the two is very small, on the order of millionths of an inch. Vibrations caused by motors, wheels, actuators, and other motions or the robot jiggling the disk head will ruin the data. In the worst cases, the disk head might touch the surface of the disk and scrape off the magnetic coating, ruining the HD completely. Read the vibration specifications of the HD very carefully before using it in a robot design. Consider replacing the HD with a more expensive alternative like flash memory cards that have no moving parts. In the design of a robot, it's wise to restrict the amount of data that will have to be stored onboard. If an HD must go into the robot, consider using an HD specifically designed for laptops. Laptop HD drives are more robust than most.

- **Shock**   If the robot hits a pothole, falls over, or simply burps, the HD may experience a sudden shock. It's not unusual for shock forces to hit 50 or 100 times the force of gravity for a very short time period. Read the HD specifications very carefully. There may be different shock specifications for operation and storage. When trying to match the HD specifications to the robot's specifications, don't forget to include the period when the robot is being shipped but is not operating. If the HD cannot take the shock specified in the robot's requirements, consider another technology like flash cards. Another option would be to consider derating the robot's specifications so it will be treated more gently in operation.

- **Temperature**   Like any component within the robot, a HD will have temperature ratings. The only extra thing to consider is that temperature might cause the HD to shrink or expand temporarily and thus make errors for a while.

- **Gyroscopic torque**   HDs contain rotating masses. As such, they will behave like a gyroscope. Remember those spinning toys that could balance on your finger? Gyroscopic action inside any HD will exert the same forces.

Any HD will resist being turned. If the robot will be spinning or moving in such a way that the HD is thrown around some, better do some worst-case calculations on the rotational velocity of the HD. The specifications of the HD may not be very clear about the torque that the operating drive can withstand without making errors. If possible, arrange for any robotic motions affecting the HD to be coplanar with the rotating mass of the HD. Picture an HD placed on a flat surface. Most HDs are shaped like a brick, with the disk surface spinning like a record rotating about a vertical axis. The operating HD can be moved sideways just fine without engendering gyroscopic forces. But if the HD is twisted sideways, it will resist with gyroscopic forces. These forces could ruin data or burn out the motor bearings over time. Read the specifications for the HD very carefully. If no specification for rotational acceleration exists, beware. All disks do have a limit for this, so find out what it is.

- **Spin-up time**   HDs take a couple of seconds to spin up to operating speeds. If the information on the HD must be instantly accessible, don't allow the HD to spin down automatically with disuse. Some computer systems will enable HDs to spin down to save power. If this is an issue, make sure the computer is not allowed to turn off the disk motor.

- **Longevity**   HDs only have finite lifetimes. There should be *Mean Time Between Failure* (MTBF) information (discussed later) for the HD, which takes certain failure mechanisms into account. Barring electrical failure, either the spindle bearing will wear out or the HD magnetic surface will wear out. In addition, some issues may arise regarding data degenerating over long time periods and becoming problematic. CDs have this problem, and HDs probably do too.

- **Error rates**   HDs do make errors. Generally, the signals that are recorded are more than sufficient to allow a proper read of the data.

  - **Bad disk surface**   HDs also have a mechanism to avoid bad spots on the disk surface. A separate place exists in the HD surface to record the location of bad sectors on the surface so they can be shipped from the factory without having perfect media. The HD can then avoid those bad spots altogether. However, the disk surface can also develop new bad spots. If the operating system can detect such an occurrence, it can compensate for it.

  - **Bad write or read**   Occasionally, the disk just makes a mistake. Errors can occur because of media problems, vibration, and probably phases of the moon! Usually, the operating system software is programmed to detect the problem and compensate for it. If the design of the robot is such that data must never be lost, then multiple disk images should be written. It is also possible to put in multiple disks to accomplish this. Read up on *Redundant Array of Inexpensive Disks* (RAID) systems if this is a requirement for the design of the robot. Check out www.acnc.com/04_01_00.html and www.pdl.cs.cmu.edu/RAIDtutorial/ for RAID technology info. For both URLs, follow the links; they have multiple parts.

- **Removable media**   The robot design may call for the addition of removable media. Floppy disks, CD-ROMS, or memory cards can be used to add or remove data from the robot. This sort of communication link works if the data does not have to go through immediately. It's called Sneaker Net because the operator walks around and carries the data. The only extra considerations to keep in mind are
  - Removable media may be less reliable than permanent media.
  - Removable media can be stolen or misplaced.
  - Removable media can jar loose with shock.
  - Removable media drives leave an extra hole in the side of the robot that can admit dirt and RF interference (to be discussed later).
- **Connections and cable integrity**   HDs have many connections. In any portable robot application, connections can be an unreliable component. If the HD connects directly to a robot PCB, then the number of connections is minimized. If the HD is connected to the robot using a flexible cable, then the number of connections is doubled (decreasing reliability) and another component is introduced into the system. A flexible cable will truly live up to its name; it will flex. If the robot moves about quite a bit, all components will be subject to accelerative forces. Most flexible cables are not meant to withstand continuous flexing. They are only made flexible so the cable can bend around and make a proper connection. Specially made flexible cables can be designed for repeated flexing in mobile applications, but they must be specified for such use, and most are not. Flexible cables may ultimately break.

Tape Drives

It seems clear that all the considerations we've discussed about HDs also apply to tape drives. To reiterate these considerations, let's list the topics from the previous discussion:

- Vibration
- Shock
- Temperature
- Gyroscopic torque
- Spin-up time
- Longevity
- Error rates

- Removable media
- Connections and cable integrity

The following issues are a bit different:

- Most tape drives that support removable media have large openings. These openings can admit contaminants and allow RF emissions both in and out of the robot.
- Be sure to check on the orientations that the tape drive can operate in. Some tape drives may have more restrictions than others.

Printers

Many of the same issues relevant to HDs are also relevant to printers. The same issues are listed here as a reminder. Some of them have been modified. If the robot must have a printer onboard, consider all the same issues.

- **Vibration**   Printers not only may malfunction in the presence of vibration and shock, but they also generate vibration and shock and distribute it throughout the robot. These extra motions must be added to the environmental specifications that the robot must withstand. Said another way, the printer's motion makes the overall vibration and shock requirements tighter.
- **Shock**
- **Temperature**   When evaluating the performance of the robot in temperature extremes, do not forget the properties of both the ink and the paper. These components must also pass muster. Don't forget that paper is a major source of dust.
- **Gyroscopic torque**
- **Spin-up time**   Many printers take quite a while to warm up, longer than disks do.
- **Longevity**   Printers are less reliable than HDs are. They also wear out faster.
- **Error rates**
- **Removable media**   Printers are likely to leave huge openings in the sides of a robot. These openings generate RF emissions and admit dust and dirt.
- **Connections and cable integrity**

Displays

When considering large displays, such as LCDs for a robot design, we have to consider multiple problems:

- **Shock**   Glass displays will crack once G forces get too high. It's very easy to break LCDs; people break their portable PC screens all the time.
- **Temperature**   LCDs have a very limited temperature range and will become unreadable outside that temperature range. Look for other display types if the robot needs to function over a wide temperature range.
- **Longevity**   The backlighting of LCDs can wear out over time, making them dimmer and nonuniform. This can happen if the LCD contains bulbs, which will suffer from metal migration over time. Look for LCDs that have backlighting using other technologies and check the MTBF specifications.
- **Power**   The LCD backlighting can take quite a bit of energy. Further, the software to control the energy expenditure can be complex to write.
- *Electromagnetic interference* **(EMI)**   When considered in terms of electromagnetic shield integrity, LCD screens are just massive holes in the package of the robot. This makes it very hard to keep radiation from leaving or entering the robot.

Process for Choosing a Robot's Computer Hardware

At some point, we'll be faced with the task of picking the actual computer to put inside the robot. This is a task that requires experience and should be approached in a systematic way. That said, almost everyone does this a different way, so have patience with differing opinions. And try to bear in mind that a diverse gene pool actually is a good idea!

The first thing to consider is getting some help. Just as advisors can be of considerable assistance in the planning phases of a project, so too can they be of value in the early *High-Level Design* (HLD) phase. The best advisors to approach are experienced engineers who have done it all before. They can often see clear solutions among the many options available to the project.

Barring the discovery of an immediate solution for the needs of the project, it makes sense to list the viable candidates that will be considered. Whether it makes sense to build or buy a computer for the robot, the end result will still be the same. The robot will have a computer board (or boards) inside it, and the boards will contain a processor chip. In listing the candidates, it makes sense to list them all as processors. Magazines list processor chips in articles that are updated at least yearly.

Here are a couple of recent web sites with lists of embedded processors that could be considered. The lists are neither exhaustive nor are they correct, but it's a place to start:

- www.e-insite.net/ednmag/index.asp?layout=article&stt=000&articleid= CA245647&pubdate=10/3/2002
- www.cera2.com/micr/index.htm

We should observe certain rules during the selection process:

- Make sure none of the requirements of the design are too restrictive. If the design requirements are so tight as to eliminate all but one candidate, then change the design. If the design or components have no wiggle room, the project is much more likely to run into problems.
- Make sure that any processor candidate exceeds most of the requirements by a good margin.
- Don't spend a massive amount of time on the selection process. Among the final candidates, more than one perfectly good choice should be available.
- Prioritize the requirements. No processor will be a perfect fit for the requirements. If the most important requirements are identified, it may help to make the choice clear.

Several possible strategies can be used for any search. Here are a few pointers that can be used during the process:

- If the project advisors suggest a specific candidate, consider it for final selection and require the other candidates to knock it out of the number one spot by exhibiting clearly superior characteristics.
- Another alternative is simply to disqualify candidates until few are left. The search process itself involves solving a number of problems simultaneously. The specifications for the robot and for the control algorithms impose many constraints on the processor selection process. Each constraint can be used to eliminate many candidates.
- Consider starting with the constraint that is the hardest to overcome. Often, the most difficult constraint will eliminate most of the processor candidates. The quicker we can narrow the field down to a very few candidates, the less work the selection process will take up.

CONSTRAINTS

What are the sorts of constraints commonly considered during the selection of a processor? The following pages contain a basic checklist of things to look into when selecting a computer.

Compute Crunch

It is quite difficult to get a sense of the amount of work the processor must perform in real time. The processor will be executing instructions at full speed unless it is asked to rest as part of saving power. During the time it is running instructions, it can only perform a discrete number of calculations per second. The amount of work the processor can do is therefore limited. It is a fine art to be able to estimate how much work any one processor can accomplish. The following should be considered during this process.

Algorithms   Perform an analysis of the control algorithms and determine how many instructions per second must be executed to accomplish the work required. If, for instance, we know that the algorithm requires the processor to perform 5,000 instructions every 20 milliseconds, we have a measure of how busy it will be. It will be executing at least 250,000 instructions per second (5000/.020). If the processor is capable of executing 1 million instructions per second, then about 25 percent of its time will be devoted to the algorithm in question.

Next, we can take a look at all the other algorithms the processor must execute and determine how much of the processor time they will consume. Simply add up the percentages to get a gauge of how busy the processor will be. As a general rule of thumb, don't count on being able to load the processor down more than 50 percent. All processors will have housekeeping tasks to perform that cannot be easily accounted for.

Further, the more often any one algorithm is executed, the more overhead it will require. Simplifying matters somewhat, it can be stated that a fixed amount of processor time is required to execute any one algorithm once. If we make the algorithm shorter but have to execute it more often, the processor will lose efficiency. Let's look at an extreme example.

Suppose, for example, that the processor must execute 100,000 addition instructions once a second. Suppose further that it requires 10 instructions just to kick off an algorithm.

Algorithm A executes 100,000 additions at the same time. To accomplish the required 100,000 additions per second using algorithm A, the processor must execute 100,010 instructions per second $(1 \times (100,000 + 10))$.

Algorithm B executes 100 additions every millisecond. To accomplish the required 100,000 additions per second using algorithm B, the processor must execute 110,000 instructions per second $(1000 \times (100 + 10))$.

Algorithm A will be much more efficient of the processor's time, about 10 percent more efficient than algorithm B. We'll study the various methods of setting up control algorithms in a later chapter.

Operations per Second   Most computers are capable of executing many instructions every second. Often, this number is given in *millions of instructions per second* (MIPS).

When evaluating a processor, it can be difficult to determine just how many instructions it truly can execute per second. This can be accomplished in a few different ways:

- **Calculated speed**   Processors have a definitive clock speed that can be determined by an inspection of the PCB circuit. Usually, there will be a "crystal clock" in a metal can immediately adjacent to the processor chip. The frequency marked on that crystal is a starting point; it's often about 50 MHz (50,000,000). It represents the external clock frequency of the processor. Often, the processor will multiply that external clock frequency by a factor to determine the internal clock frequency. Once we know the internal clock frequency, we need only determine the number of internal clock cycles that the average instruction consumes. Most processors fix this number between 1 and 8. With these numbers in hand, we compute the number of MIPS for the processor on that PCB.

- **Benchmarks**   Many people are curious about the speed and capability of different processors and they're eager to see comparisons. One of the difficulties involved in making such comparisons is that each processor has its own strengths and weaknesses. Even a wimpy 8-bit machine with the right instruction set can whip a 32-bit machine in some tasks. It depends on what the processor designs were optimized for.

  Various organizations have proposed standardized programs (called *benchmarks*) that can be executed on every machine. Generally, the benchmarks are simple programs for sorting, moving, or transforming data. Invariably, the benchmark then rates the performance of the processor with some index, which can be compared to that of the other processors. Since all the benchmark programs are different, it is not unusual for processors to be rated at the top of a benchmark, but lower on others.

  If we want to determine from benchmark data how processors will do in our robot design, we must find a benchmark that performs a task similar to the algorithms in our robot design. If we choose a benchmark that performs tasks sufficiently different from our robot, we may be led astray. Generally, benchmarks are available for downloading and execution, but they can take quite a bit of work to implement. Here's one (older) PDF file describing benchmark techniques. The data may be obsolete, but the methods are still fairly modern: www.zilog.com/docs/z380/z380 bench.pdf.

  These sites provide descriptions of a few benchmarks:
  - www.specbench.org/osg/cpu95/news/cpu95descr.html
  - http://spec.unipv.it/
  - http://www.eembc.org/
  - http://performance.netlib.org/performance/html/dhrystone.intro.html

- www.dl.ac.uk/TCSC/disco/Benchmarks/whetstone.html
- www.dl.ac.uk/TCSC/disco/Benchmarks/spec.html

The following URLs contain some benchmark results:

- http://spec.unipv.it/results.html
- www.eembc.org/
- http://kennedyp.iccom.com/riscscore.htm
- http://kennedyp.iccom.com/cpuscore.htm
- www.cpuscorecard.com/benchmarks2.htm
- www.netlib.org/performance/

So what do we do with all these numbers? We've seen how we can get a parametric evaluation of a processor. If we have done our calculations properly (in the case of calculating MIPS) or if all of our assumptions are correct (about the applicability of benchmark data), then we can compare processors directly in terms of compute crunch. The processor with the best performance can be chosen (all other things being equal).

Arithmetic Capability

Most processors have fixed-point instructions. Some processors have floating-point arithmetic instructions built in. The PowerPC™ is in this class (see http://developer. apple.com/techpubs/mac/PPCNumerics/PPCNumerics-146.html).

If the robot must process quantities of floating-point numbers, this capability will be important. Processors with just fixed-point instructions can still execute floating-point instructions, but the execution will be much slower than that of a processor with intrinsic floating-point instructions.

Word Length

We've discussed word length before, but it's worth listing again. It's certainly an important characteristic of a processor. Processors with longer word lengths generally have added capabilities that make them much faster than one might expect. A 32-bit processor is generally much more than 4 times faster than an 8-bit processor.

Memory Size

Many small processors have RAM and ROM memory built right into the processor chip. It's easy to get 8-bit processors with such internal memory, and even some 32-bit processors. If the robot's control program is very small, and the number of robots to be built is large, consider these sorts of chips. It's wise to double all estimates of memory size; get much more than might be needed. The larger the memory size, the better.

The size and type of cache memory is also an important feature of a processor. If the robot's software has many tight execution loops, then cache can be a very important feature to have in the selected processor.

Bandwidth

We've already discussed bandwidth, and it belongs on this checklist. Be sure to look into all the different bandwidth limitations of the processor.

Utilization of Resources

In many real-time applications like robots, various processor resources are extensively used. Very often, as multiple control algorithms execute, they have to share resources inside the processor. If one of the resources is in great demand, it can become the bottleneck in the system that limits overall performance. The DMA capability of the processor is a typical example. Some smaller processors have a special-purpose memory controller (DMA) that can move blocks of memory around. It works in parallel with the processor when it's started. If it cannot get its job completed by the next time it's needed, then it's clearly overused. If an inspection of the architecture points to such a problem in advance, then look for a processor with more capabilities.

Special-Purpose Hardware

We discussed several special-purpose processor types previously, and it's worth mentioning them again on this checklist. Many processors, even general-purpose processors, have one or two special features worth noting.

Power

Power considerations are one of the most critical features in a robot and in a processor. Later, when we talk about power control, we'll get deeper into the details. If the robot is battery powered, then give considerable attention to power matters. The success of power conservation efforts very often hinges on the power-saving capability of the processor and the software available for using its power-saving features.

Cooling

Even if the robot can supply lots of power to a processor, we must take into account the method of cooling the processor and circuitry. Some processors require heat sinks and even fans, both of which can cause reliability problems and take up space.

Voltage

Processors are available that work on various different voltages. Some processors can work on very low voltages, as low as 1 volt (although this is rare). Moving to a lower-voltage operation can be done for two reasons. The first is power savings, which we'll discuss shortly. The second reason is to match the battery voltage (if a battery is used in the system). A considerable amount of power-supply circuitry can be taken out of the design if the circuitry can accept the battery voltage directly.

Space

Sometimes space is at a premium inside a robot. If so, consider the space taken up by the processor chip and any heat sinks or fans that may be required.

Reliability

Some robots are sent to space or other relatively inaccessible places. They may be subject to extreme environmental stress, including temperature, vacuum, vibration, and radiation. There may be no opportunity to even service the robot. If reliability is key, then consider the choice of processors carefully. Spacecraft designers, for instance, often choose older, proven processors that have been tested for years. There are several good articles on the special considerations must be taken into account:

- www.klabs.org/DEI/Processor/index.htm
- http://klabs.org/DEI/References/design_guidelines/design_series/1248.pdf
- www.gd-ais.com/Products/srs/process/isc.pdf
- www.spacecoretech.org/coretech2000/Presentations/Software/ISC_Case_Study/sld001.htm

Reprogramming

Some computers have onboard memory. Make sure to check if this can be reprogrammed or not. Even if it can be reprogrammed, check and see which features the processor has to facilitate it. How will the processor be accessed, downloaded, and restarted?

Benchmarks

We've discussed benchmarks and how to determine the "horsepower" of a processor. It makes sense to list benchmarks on this checklist.

Price

It's possible to obtain processors for under a dollar, and it's possible to pay thousands of dollars for some. When shopping for processors, don't forget that a quote means nothing if delivery is not forthcoming. Some processor companies will not even talk to customers who only want a few processors.

Software Tools

A good processor that is not well supported is useless. Here are some factors to consider:

- **Support**   How well is the processor supported with software development tools? Many special-purpose processors have custom-made tools that must be used. Most general-purpose processors attract sufficient attention so that multiple software tools are available.
- **Second source**   What will happen if the processor company goes out of business? It is possible to specify processors for the robot that can be purchased from multiple vendors. The processors may not be identical in all cases, but the conversion will be simple. Don't forget that it does no good to secure two distributors for a processor if the processor is only made in one place!
- **Availability of SW engineers**   Don't forget to consider the software engineers when choosing a processor for the robot. If the SW engineer cannot handle the chosen processor, consider replacing the processor. The other, easier option is to replace the programmer!
- **Software price**   Don't forget to consider the price of the software tools. Many are available on the Internet, but others can be very expensive.
- **Development equipment**   Some software tools may require a very fast development system, and it may not be a PC. If the development system requires Unix or another operating system, a PC may not be the right choice for the hardware platform.
- **Licenses needed**   Consider how many programmers must use the development software tools. Some tools come with "seat" licenses, meaning the project will need one license for each engineer using the development tools.

Development Time and Expense

Beyond the issue of software development tools, many other issues can affect the time and expense of the development of the robot:

■ **COTS**   This acronym stands for *Commercial off the Shelf*. Basically, we can buy computer boards with processors already on them. We've discussed this before, but it bears listing here. For each processor candidate, how many companies are selling PCBs (suitable for the robot) containing the processor?

■ **Third-party software**   Has anyone written specialized software for this processor that can be of use in the robot design?

■ **Tech risk**   Will any features of the processor help defuse the project's technical risk?

More issues should be considered when choosing a processor. Here are a few more web sites and PDFs detailing how others might approach the process:

■ www.cs.berkeley.edu/~liblit/darwin/slides
■ www.cs.berkeley.edu/~liblit/darwin/darwin.pdf
■ http://bwrc.eecs.berkeley.edu/Publications/2000/Theses/Evaluate_guide_ process_archit_select/Dissertation.Ghazal.pdf
■ http://dec.bournemouth.ac.uk/staff/awatson/micro/notes/Lect1_98.doc

Picking a processor and computer hardware for the robot is much more complex than it may seem at first glance. But with proper attention paid to all the questions outlined above, the process should go smoothly.

RELIABILITY, SAFETY, AND COMPLIANCE

Reliability

Why bother with this topic at all? Well, given the most recent TV shows about battling robots, most people think of robots as mechanical disasters that can only last three minutes before various parts start to fall off (or get yanked off). Whole organizations are devoted to such events, such as the Survival Research Labs (www.srl.org/). On the more serious side, robots working in automotive plants are expected to work nonstop for years in very difficult conditions.

To the greatest extent possible, it makes sense to design a robot to be highly reliable. Towards that end, we must learn what reliability is, how to measure it, how to predict it, and how to achieve it. Certainly, many ways are available for accomplishing this, both scientific and seat of the pants. In this chapter, we'll take a grand tour of both methods.

Reliability has many definitions; here's mine. For the robot to be reliable, it must fulfill all of our expectations. Certainly, the tires cannot fall off. But more to the point, it must be capable of achieving the goals that we set for it. These goals may include performance, production, uptime, and dependability. Bottom line, we will rely on the robot and it has to come through.

The *Institute of Electrical and Electronics Engineers* (IEEE) defines reliability as "the ability of a system or component to perform its required functions under stated conditions for a specified period of time" [IEEE 90]. Further definitions can be found at http://athos.rutgers.edu/~rmartin/teaching/fall99/lectures/10/gfx004.html.

By observation, engineers have documented the failure rates of various component types. Bellcore, now called Telcordia, has documented many of these failure rates and published them at www.t-cubed.com/faq_bell.htm.

MATHEMATICS

Let's call the failure rate of a component λ, measured in failures per unit time. Loosely speaking, if λ is .001 per year, we could expect an average of one failure per year in a population of 1,000 such components. We further define *Mean Time to Failure* (MTTF), as the inverse of λ. It represents the average amount of time a single component is likely to last before it fails the first time.

$$MTTF \ = \ 1/\lambda$$

We adopt MTTF as a convenient way of measuring and doing calculations about reliability. However, some limitations exist, as explained at www.reliasoft.com/newsletter/2Q2000/mttf.htm.

Once we accept MTTF and λ as viable metrics, they can be used in calculations in the following ways:

- Clearly, the reliability of a component can be defined from either MTTF or λ, since they are the computational inverse of one another.
- If a system has multiple components, then λ of the combined population is the sum of the individual λs. $\lambda pop = \lambda 1 + \lambda 2 + \lambda 3 + \ldots + \lambda n$. Effectively, the failure rates add up. If the components are all on a *printed circuit board* (PCB), for instance, then the failure rate of the PCB is the sum of all the failure rates of the individual components. Clearly, the PCB will be less reliable than any one component, and since the least reliable components have the highest λ values, they may well determine the overall reliability of the PCB.

  Note that a combined population of even the most reliable components may not be reliable. The chance of having one failure in the population may be high if many individual components exist, even if they all have a low failure rate. Said another way, since $\lambda pop = n \times \lambda$, if n is large, λpop may be large even if λ is small.

- If we combine two systems with MTTF1 and MTTF2, we can find $\lambda 1$ and $\lambda 2$ by inversion, add them together, invert the result, and get the MTTF for the combined system. In this way, the reliability of a large system can be predicted.
- To predict the MTTF of a large PCB, for instance, perform the following steps:
  1. Obtain the *Bill of Material* (BOM) for the board, the list of parts, and the quantity of each.
  2. Make a spreadsheet of the part types, grouping similar parts together. All the resistors of the same wattage, for instance, can be thrown into the same group.
  3. Don't forget to count components like feed-throughs on the PCB. They have failure rates, too! Count all connections and wires.
  4. List the quantities in each group.
  5. In the Bellcore tables, look up the λ failure rates for each component type.
  6. Multiply the failure rate by the quantity for each component-type failure rate, and add up the component-type failure rates to get a failure rate for the PCB. Invert it to obtain the PCB's MTTF.

MTTR is the *Mean Time to Repair a Failure*. Basically, it's the average time it takes to repair a failed component and bring it back to functioning status.

MTBF is the *Mean Time Between Failures*. It's related to the MTTF and MTTR in that it is the sum of the two (MTBF = MTTF + MTTR). Basically, once a component is repaired after a failure, it will take an average of MTTF + MTTR for it to fail again and be repaired. (For more info, go to www.ab.com/harry/mtbf.html or www.its .bldrdoc.gov/fs-1037/dir-022/_3254.htm.)

AVAILABILITY

The reliability of many systems is measured as availability. People who use computer systems get very impatient if the computer goes down and cannot be used. They are specifically interested in the percentage of time that the computer system will be usable. Thus, availability is defined as follows:

$$Availability \ = \ MTTF/(MTTF \ + \ MTTR) \ = \ MTTF/MTBF$$

From the middle term in the equation, we can see that if MTTR is very short, or if the MTTF is very long, then availability approaches 100 percent. Engineers who need high availability can work on both the MTTF and MTTR to achieve their goals. The MTTR can be lowered in several ways.

If a failed unit requires a trip to the repair depot, it can take quite a while. But if a spare component is right on site, repair can take just a few minutes.

If a hot backup component can switch over automatically, it will take a few seconds. Hot backups can be accomplished with what's called an "N + 1" backup. If a total of

N components are necessary for the successful operation of the system, the designers add one more system, hence the terminology N + 1. If one of the N components fails, the spare component will take up the slack and the overall system will keep running. If, however, a second component fails before the broken one is repaired, then the system fails. The calculations for MTTF and availability are somewhat complex in a case such as this. For further information, see the following sites:

- www.mapleapps.com/categories/engineering/manufacturing/html/reliability.html
- www.mathpages.com/home/kmath326.htm
- www.mathpages.com/home/kmath498.htm

Done properly, availability goes up quite high, but at the price of a single extra component. Repair is basically instantaneous, except for the case where two or more components fail at the same time. It's therefore wise to make the repair to broken units as rapidly as possible. This, too, increases the availability.

COMPONENTS

Another way to increase the MTTF of a system is to avoid unreliable components. The Bellcore component tables list the failure rates of many different types of components. Without the tables in front of me, I'll list the components I remember as having high failure rates. These are components we should avoid in the design and construction of the robot.

Connectors

Connectors of all sorts are lower-reliability parts. Every single pin on every connector must be counted. Connectors generally work by spring pressure. A bent piece of metal pushes on another piece of metal and thus excludes gas and dirt. Often, a wiping action is made as the connection is made. The following problems can make connectors fail.

- **Contaminants** If we have too many contaminants or gases are not excluded, corrosion can creep into the connection and ruin it.
- **Currents** In addition, if high currents move over the contacts, they can corrode from too much heat.
- **Vibration** Movement can cause contacts to break open, interrupting signals.
- **Operator error** Service technicians can fail to seat connectors properly. Stating a cynical view, the function of a switch or connector is to be installed wrong or set to the wrong position. If connectors are not used, they cannot present such problems.

- **Shock**   Sway and shock can tear wires out of connectors.
- **Bad design**   Contacts are one of the hardest components for an electrical engineer to master. They always take contacts for granted. The truth is, contacts require quite a bit of experience to use correctly. If a designer puts too much current through them, they can burn out. If the designer does not have sufficient voltage across them, contacts can fail to make true contact. The failures occur well after shipment.

Cabling

Wires and cables can fail for several reasons listed below.

- **Abrasion**   Wires and cables will rub up against surfaces and abrade the insulation away. In a robot, this could be a real problem. Cables should be well insulated and run far away from sharp internal surfaces. Most PCBs are full of sharp objects.
- **Flexing**   Wires and cables will flex repeatedly inside a moving robot. Eventually, one or more wires may break. Flexible cables are a discipline unto themselves in the electronics world. It takes great experience to build them properly. If a cable is not designed to flex repeatedly, see to it that it does not!

Transistors

Even though a processor may contain 4 million transistors, it may be more reliable than a single transistor! In particular, power transistors can be a problem. Many engineers are not well versed in keeping them safe and happy. As a general rule, try to avoid using too many discrete transistors. If the design calls for a power transistor, make sure it's used well within its specifications.

Batteries

Batteries are basically canisters of chemical soup destined to leak, die, explode, and fail. If we are lucky, they will not do all of these things at once. We'll get more into safety later. Most robots probably will have batteries in them. If batteries are going to be in the robot, better study the technologies very well. If the batteries can be replaced periodically, preventive maintenance might even be warranted. Some batteries are more reliable than others. If the robot is designed for long-term autonomous operation, then study spacecraft technology and the batteries used in satellites.

Bulbs

Bulbs with filaments are quite fragile and will fail quicker than most. Even if the robot does not move, bulbs are likely to fail at a rate similar to those in households. Gaseous bulbs will also fail over time, much like overhead fluorescent bulbs do. Consider white *light-emitting diodes* (LEDs) for illumination in a robot design. Many LEDs are specifically designed to provide usable reading light.

Moving Parts

Moving parts are particularly subject to wear. Further, as they wear out, they may shed debris. Lubricants, like oil, have to be used with extreme caution. They can easily migrate to connectors and cause contact failure. Consider life testing any moving part to a number of repetitions well beyond its expected lifetime count. Life testing involves subjecting the relevant components with conditions that simulate years of use in a much shorter test time. The following web site has a very nice treatment of reliability issues: www.itl.nist.gov/div898/handbook/apr/apr.htm.

Safety

If the auto industry were like the computer industry, a car would now cost $5, would get 5,000 miles to the gallon, and at random times would explode, killing all its passengers.

— *John Chambers*

Safety issues can be divided into a few different categories. Human safety is paramount. I suppose that's why they started shooting chimps into space first! After human safety, we consider the safety of the mission, so it runs to completion.

HUMAN SAFETY

Robots come in all shapes and sizes. Some popular TV robot warriors are a couple of stories tall and spit flames. Certainly, these robots present a threat to human safety, just from their physical prowess. That said, I've opened up my son's broken Furby robot. Though furry and small, it did die an ugly death by catching fire, emitting smoke, and landing outside the house in a snowdrift. Further problems were probably forestalled because the toy was properly designed to conform to applicable safety regulations, which limit the size of such fires. And all this happened just when I was getting used to speaking Furbish with the little guy.

As an aside, groups of people are significantly worried about the very presence of robots. One such group is the Anti-Robot Militia (http://unite-and-resist.cloudmakers. org). It's the one group mentioned in this book that I find a bit disturbing for a list of reasons that include, among other things, incitement to violence. But I include it here for reasons of balance, intellectual curiosity, humor, and a minor sense of civic duty. I believe that at least some truth can be found in the utterances of all people, if one reads with judgment, care, and discerning eyes. I still don't quite know what to make of the site.

Let's get back to the topic. Many aspects of robot design can cause problems, even injury, for designers and users of robots. To avoid such problems, consider designing the robot according to published safety standards such as UL or CE. Safety agencies such as Underwriter Labs and TUV can provide written guidelines. Advice herein can provide guidance, but to be sure about the safety of the robot's design, consult these organizations. These groups publish safety regulations and offer testing services at their labs. At the very least, purchase the relevant safety specifications mentioned at www.ul. com/robot/. Further sites to check out include www.tuv.com and www.1metlab.com.

To supplement the safety recommendations from the standards agencies, here are a few more pointers. If, for any reason, information herein conflicts with information in the safety standards, follow the safety standard first and foremost.

Panic

Any robot should have a panic button that is red, visible, and intuitive to use. Make it his nose if need be, but don't forget to put one on. The button should stop all robot activity and shut down the power systems at the power source. Once somebody is scared enough to press the button, it must provide immediate relief to all concerned. That said, don't forget that every kid (big and small) will be sorely tempted to press the button. (Am I right out there, kids?) If the robot is carrying out some critical function that should not be interrupted lightly, then put some warning, or kidproof shield, over the panic button.

Batteries

As a reservoir of energy, batteries present a natural threat to humans. They come in all shapes and sizes and are sold over the counter. But the more exotic batteries, with characteristics of particular interest to robot designers, can be much more dangerous than the batteries sold in stores. They look very much the same since they are designed for the same types of battery holders. Take as a clue, for instance, the warning on many toys against using rechargeable batteries in the toy. The following sections discuss some of the things that can go wrong with batteries.

Leakage Batteries contain chemicals that can leak out and ruin portions of the robot. In the case of a car battery, for example, this would be a real possibility if the batter is turned upside down. Liquid acid will ruin anyone's day. Other batteries contain other more solid chemicals that still present leakage hazards. Treat batteries like a container of chemicals. Make provisions outside the battery to contain any leakage that might occur. If serious concerns exist about leakage, search for a type of battery that will not "leak," if such a battery exists.

Explosion Some batteries are capable of releasing their energy very rapidly. This can happen if they are shorted or if the robot's power supply shorts out. When this happens, energy is dissipated inside the battery and an explosion can ensue. If the design calls for a type of battery that can explode under these circumstances, then consider the following safety steps:

- Read and understand the manufacturer's warnings.
- Determine just how big such an explosion might be and treat the batteries as an explosive.
- Consider housing the battery in an explosion-proof container so it will not cause damage during a catastrophic event.
- Educate all the engineers about how to handle the batteries, or restrict the number of people authorized to do so.

Jewelry I include this section in addition to the following warning about fire for a very specific reason. Some batteries can release massive amounts of current in a very short time period. If a designer is wearing a ring that shorts out the battery, the ring can heat up sufficiently to take off a finger. Do *not* wear jewelry when working with batteries. For that matter, designers working with electronics should consider removing jewelry during work hours. I would include rings, watches, and necklaces in this category. Freak accidents happen much more often than expected.

Fire Just as batteries can heat up jewelry, they can also start fires inside the robot's wiring. Consider putting a suitable fuse or circuit breaker in with the batteries. If a short circuit someplace within the robot should occur, this may help. Don't forget to use UL-rated wire and PCB components that are somewhat fire resistant.

Trauma Don't drop batteries on your foot! I'm just waiting for such a warning label to show up. I think it's only a matter of time.

MECHANICAL THREATS

An interesting story comes to mind, so far uncorroborated. Coal power plants have had a problem matching their power generation to the required loads. Coal fires take a long time to burn out, so energy can be wasted in the evening hours as people retire for the night. Power companies have been looking for ways to store the wasted electrical energy so it can be reused the next morning.

Clever designers decided to investigate putting the energy into a flywheel. The flywheel was to be a large rotating wheel driven by a motor/generator. In the evening, they would speed up the flywheel to store up some of the wasted energy from the coal fires. In the morning, the flywheel would spin the generators to reclaim the energy before the coal fires were completely burning. The flywheel was to be massive, an intimidating hunk of spinning concrete and steel. Clearly, if the flywheel ever came loose, it would be quite dangerous. The design plan was to set the flywheel on the edge of Long Island Sound, a 10-mile wide body of water between Long Island and Connecticut. If the flywheel ever broke loose, it would roll into the water and thus dissipate its energy. Calculations showed it would only get three-quarters of the way across to the Connecticut shore before it rolled to a stop! I'm guessing this information was cold comfort to the beach residents in Connecticut.

In general, inspect the robot design for places where energy is stored. It does not take much energy to create injury or to cause a breakdown in operation.

The following is a list of items to check for and help avoid mechanical problems:

- **Leverage**   Even a small force can be greatly magnified with leverage. Inspect the robot design for hazards that might be created by excessive leverage or force. Such hazards probably require shields to prevent accidental injury.
- **Sharp parts**   If the robot does not require sharp-edged parts, round them off in the design phase.
- **Fast moving parts**   Even lightweight parts can cause injury if they move fast. Shielding or redesigning the relevant parts might be in order.
- **Stressed parts**   Parts put under stress might break catastrophically. Portions of the part may fly off. Examine all parts of the design for hidden stress and at least understand what happens should the part break.

SOUND PRESSURE SAFETY

The human ear can only withstand certain sound pressure levels before injury occurs to the ear. If portions of the robot are to be noisy, then either calculate the predicted noise level or measure it directly.

LASERS AND LIGHT SAFETY

Radiation, at any frequency, can damage humans. The government regulates the power that lasers can emit so it is "eye safe." Although most small pointer lasers are eye safe, it's still not a good idea to employ such radiation. Be very careful about the amount of light and laser light that the robot will emit.

FIRE AND ELECTROCUTION

UL and other companies we've discussed publish testing and design guidelines that should be followed.

Environmental Considerations

In the design of a robot, we must pay attention to a number of environmental considerations. Several factors in the robot's environment will be critical to its reliability and performance. Other factors are a bit less important. Among the factors to consider are temperature, vibration, vandalism and theft, humidity, altitude, and contaminants.

TEMPERATURE

Most electrical systems and appliances are designed to work in an office environment near 25 degrees centigrade (25C). Commercial-grade electronics are designed to work from freezing to hot (0C to 70C). Most off-the-shelf systems have temperature specifications in this range. Industrial systems, designed to be a bit more robust, have a temperature range of -40C to 85C. Military systems often have temperature specifications of -55C to 125C. Separate temperature ranges are quoted for storage and for operation. The worst-case temperatures often occur in vehicle applications, especially the very high temperatures that can occur in car trunks.

In the design of the robot, study the temperature limitations of all the components. Often the battery and displays will be the least robust parts, with a limited temperature range. It is possible to violate temperature specifications, but it would be taking chances; the system might fail. It is also possible to test the completed system at high and low temperatures to help ensure that it will be able to handle temperature extremes. If a system is exercised for about 4 days at its high temperature limit, it's roughly comparable to 30 days of aging in the field.

Most semiconductors and other parts have what is called a bathtub curve for reliability. Most of the failures occur during the first few months (infantile failures) and the last few months (end-of-life failures) in the lifetime of the robot. Hence, the failure rate curve over time looks like the outline of a bathtub. By baking the robot for a few days at an elevated temperature, many of the infantile failures can be precipitated in the manufacturing bay, instead of out in the field (see Figure 4-1).

Here are a few web sites discussing this effect:

■ www.asq-rd.org/articleBathtub.htm
■ http://ranger.uta.edu/~carroll/cse4317/reliability/sld004.htm
■ www.itl.nist.gov/div898/handbook/apr/section1/apr124.htm

VIBRATION

Vibration can tear a robot apart. The vibrations might come from an external source, like a vehicle the robot is riding in, or the vibrations might come from the robot itself. But how does one go about quantifying the threat and preparing for it?

Vibrations are generally measured in terms of accelerative forces and frequency. I will detail my personal method for dealing with this problem. These are methods that have worked for years to toughen up designs. Also, a short list of web sites covering vibration analysis is provided later for further information.

First and foremost, to really condition the robot to withstand vibrations, purchase or make a vibration table. Branford makes the table I've used before. They've since been acquired by Cougar Industries (www.cougarindustries.com/).

The table is basically a steel slab sitting on rubber pillows so it can vibrate. Bolted to the underside of the table is a motor with an offcenter, rotating weight. As the motor spins, the weight vibrates the table. The faster the motor goes, the faster the frequency of vibration. The further offcenter the weight, the stronger the vibrations are in terms

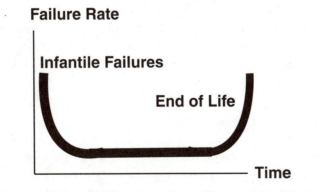

FIGURE 4-1   Component failure rates look like a bathtub curve.

of amplitude. Be aware that such tables have a weight limit. The robot must be light enough that the table can accommodate it and still be able to vibrate. The vibrations are effectively three-dimensional with each dimension's vibration approximated by a sine function. The amplitude of the vibration is as follows:

$$x = K \times sin(\omega \times t)$$

The accelerative, second derivative is $d^2x/dt^2 = -K \times \sin(\omega \times t)$. It's easy to measure the vibration forces with a strobe light. Strobe lights can be found at these sites:

- www.aaroncake.net/circuits/strobe2.htm
- www26.brinkster.com/strobeit/
- www.cpcares.com/9940.html
- www.123dj.com/l_strobelights.html

Decide how fast to vibrate the robot during the test. Obtain a strobe light and increase the strobe frequency until it matches the frequency of the vibration. The frequency of the strobe will give ω. The amplitude of the vibration can be measured with a ruler as the vibration slowly moves back and forth (looking like it's almost standing still) in the strobe light. Using these methods, we can determine both K and ω. The computed acceleration from the second derivative above can be converted to G forces, the most common method of specifying vibration force. Using these methods, I routinely test products for up to a minute at 10 Gs of vibration force at 10 Hz. Components that react badly to vibrations will appear to sway more in the strobe light than the rest will. Corrections can then be made to their mountings.

The following URLs have further information about vibration analysis:

- www.cage.curtin.edu.au/mechanical/info/vibrations/
- www.cage.curtin.edu.au/mechanical/info/vibrations/tutor.htm
- www.mech.uwa.edu.au/bjs/Vibration/default.html

Believe it or not, I have always added one extra extreme test during the design of a product. Put the product into its shipping container and drop it repeatedly from a height of three feet. Then roll it end over end down the floor for 100 feet. Open the package, look for damage, retest the robot, and change the design to fix any weaknesses that are revealed.

VANDALISM AND THEFT

I worked with a large company that prided itself on the design and manufacture of lighting systems. These systems would go into large, big-city high schools' auditoriums to control the lights on stage. A new design was being tested when we realized that the high school kids were chewing the buttons off the console! So be wary; a robot is certainly a tempting target for curious people. It *will* suffer damage from playful and malicious people alike. It may fall into the hands of operators with less than good intentions. Prepare for it.

HUMIDITY

Designing a robot for humid environments can be quite difficult. Review all the specifications for all the parts. Rust and mildew can certainly become problems. In some cases, condensation can form and short things out.

ALTITUDE

Batteries and LCD displays can become problems at high altitudes. Read the specifications for all the components to set the altitude limits for the robot.

CONTAMINANTS

If the robot must function for long time periods (years) or will be subject to a polluted atmosphere, consider an accelerated test for corrosion. The *Network Equipment Building System* (NEBS) standards were written to help guarantee the reliability of phone switching equipment at the phone company. One of the tests involved subjecting a system to a chemical fog for a week or two. The concentration of chemicals is sufficient to simulate years of operation in a poor environment. Contact Metlabs or others to inquire about the testing regimen (www.metlab.com, www.metlab.com/pages/nebs.html).

Common Sense

Many design rules for robots (or other complex systems) come naturally with experience. Here are a few words of basic advice.

COMPLEXITY

We've talked about it before. Keep things simple. Instinct should tell us if things are too complicated. Chop the robot down to size periodically during the early design phases. Take stuff out, eliminate actions, remove conditions, and take heed if people wrinkle their brow when they hear how things are supposed to work.

COMPARABLE SYSTEMS

During the design of the robot, look for comparable designs. Others have already designed many of the subsystems in one fashion or another. If the design of the robot is a significant departure from standard designs, then conduct more reviews of the design. With luck, new ground is being broken. Without luck, a disaster might be in the making.

ACID TEST

When I was at college at Cornell, a rumor was going around, the very sort of thing spread by young college kids with little experience. To this day, I'm not sure if it's true or not. It seems I have gained little sense since! The campus has two deep gorges, one spanned by a tiny suspension bridge that wobbles as it's crossed. It was said that the architecture school conducted a design contest for the bridge. The student with the winning design, returned and was surprised to see his design spanning the gorge, and refused to cross the bridge!

We must be willing to be cradled in the metal arms of our creation. If we tremble at the thought, we should review our designs!

PLAN ON FAILING

Face it; nothing works as planned. Unforeseen circumstances always take place in life and in projects. The prudent thing to do is to plan for recovery while standing amid the ashes of failure. A couple of precautions can be taken.

Watchdog

Most complex computerized systems will just plain fail now and then. The reasons may never even be discovered. It is often helpful to design a "watchdog" circuit that can reset the computer system or restart the robot if it fails to regularly interrogate the watchdog. The existence of a watchdog circuit generally increases the availability of the robot and only rarely interferes unnecessarily.

Backup Plans

We might as well plan for portions of the robot to fail. If the robot is to be autonomous, or in a position where it cannot be repaired, then special attention should be paid to backup systems. We already talked about N + 1 redundant systems, but other options are available. Consider the recent success of an interplanetary probe. The main communications antenna failed to operate, but it had a slower backup radio system. The ground controllers cannibalized part of the bandwidth of the backup radio to send the mission data back to Earth. The mission was still a success.

TESTING

Through testing, it's possible to decrease the likelihood that the robot will fail. We can gain confidence in the capability of the robot to function under adverse conditions. Further, through stressing the robot, we can precipitate failures that might occur early in its "career." Semiconductor makers, in an effort to produce more robust products, routinely test their integrated circuits (chips) before shipping them. The chips that are not tested for temperature go out with the commercial temperature rating. The chips that are tested at higher temperatures get the industrial and military temperature ratings. Some manufacturers may sample test batches of chips to estimate the performance of the entire chip population. This is a valid technique but would be useful for us only if the population of robots was large.

We've already talked about temperature testing and vibration testing. Each can be used to increase the reliability of the robot prior to use. These techniques can be used in the production of robots, but they are also of great use during the design of the robot. Weaknesses in the design will become apparent; they can be fixed prior to further development.

A further testing technique, usable during development, is more subtle. As a robot designer, don't forget that others will see the robot in a much different light. Never underestimate the ability of a three year old to walk up to the robot and say, "What happens if I do . . . *this*?" Thus will be discovered a catastrophic weakness in the design that has been sitting right on the surface unobserved.

Beyond such dramatic tests, consider putting together a series of alpha and beta tests if the robot is to be manufactured. The definitions of these terms vary, but I can outline mine.

Alpha testing is a situation where we give the robot prototype, or the initial few production units, to friendly end-users who can use it and provide constructive criticism. Alpha testing is a time period where distribution is very limited, failures are expected, and corrections are still being made.

Beta testing is when production robots are sold in limited quantities to end-users. The goal is to see just how things go before jumping into full production. The end-users may or may not be aware the units are being beta tested. At the end of beta testing, some corrections can be made, and full production and distribution ensues. Consult the following web site to learn more about the process of testing: www.cs.berkeley.edu/~jasonh/presentations/SoftwareTesting-cs169-nov1998/.

Emissions

Cells phones often drop out or have significant static on the line in the presence of interfering appliances like computers. Car radios often buzz when we drive under power lines. Computers can bomb completely if they get hit with a big static spark. These occurrences are caused by electrical interference from outside the appliance. Thus, two goals for the design of the robot spring to mind:

- We should make the robot impervious to interference. This way, it will be more reliable.
- To be a good electric neighbor, we should design the robot so it does not create interference that will be picked up by other appliances.

For reasons of symmetry, it turns out that these two goals are one and the same. If we can keep interference from entering the robot, then interference cannot get out of the robot either. To accomplish the goals, we will employ two basic methods:

- **Generation** We will try not to generate interference within the robot. If we minimize the interference we generate, we will not have to struggle to keep it within the robot.
- **Shielding** We will try to put up sufficient shielding around the robot to help prevent our interference from getting out. Further, these shields will help keep outside interference from getting in.

As a practical matter, we cannot be perfect in either endeavor. The robot will generate interference, and it will spread beyond the walls of the robot. We can use many techniques to minimize interference and accomplish our goals.

GENERATION

To appreciate just what interference is, we should go back to the works of the master. In 1873, James Clark Maxwell (see Figure 4-2) set out the very basic laws of physics in his publication *A Treatise on Electricity and Magnetism*, including the formulas known as Maxwell's Equations (for more info, access www-gap.dcs.st-and.ac.uk/~history/Mathematicians/Maxwell.html).

The presence and movement of electrons creates electrostatic and electromagnetic fields. These fields create action over a distance. A magnet, driven by force, near a wire can move electrons in the wire to create a current (creating a generator). A current, moving in a wire near a magnet, can create force on the magnet (creating a motor). In both cases, the fields involved are acting over a distance. So, too, electrons moving within

FIGURE 4-2 James Clark Maxwell

the robot can affect other electrons over a distance, the popular notion of interference. Interference coming out of the robot can be measured with antennae outside the robot.

To illustrate this effect, take an AM radio and tune it between stations, where only static is heard. Turn up the volume and put the radio down next to the computer. Execute a few computer programs to exercise the computer. The inner workings of the computer will be audible on the radio!

So how do we prevent the generation of interference inside the robot? First of all, we cannot prevent it completely. All electrical systems generate interference. The trick is to keep it well below the tolerable levels prescribed by the governmental groups that regulate it. The *Federal Communications Commission* (FCC) does this, and many foreign governments enforce the CE mark overseas. Many techniques are available for limiting the amount of electrical emissions generated within a robot.

Use Low Frequencies

All electrical signals emit interference, but lower-frequency signals tend to emit less. Further, the FCC is more worried about higher frequencies than lower. As an example of what can be done, some computers are optimized to run at clock frequencies of 32 kHz. This is a much slower clock than most computers have. As long as the computer is fast enough to accomplish its work, such a clock speed will suffice. Don't run the computer in the robot at clock speeds that are greatly faster than needed.

Use Long Rise-Time Signals

No signals inside the robot really change from low to high voltage instantaneously. If they did, the emissions would be of unlimited frequency. As a practical matter, signals rise over a certain time period; let's call it T. When this happens, emissions centered at a frequency of about 3/T predominate in the emissions spectrum. Generally, FCC regulations restrict higher-frequency emissions to lower values, so it makes sense to limit the rise times of signals within the robot, which can be done in several ways:

- First, we can use integrated circuits that have slower signal rise times. Use the chip technology that is just fast enough for the robot, but not much faster.
- Use lower clock frequencies. Although this does not guarantee slower rise times, it often helps.
- Attenuate the signals with filtering components. Electrical engineers often must alter signals so they will not generate transients on a PCB. Transients on one signal can create errors on that signal or on other signals. As the transients are attenuated, so too are the high-frequency components of the transients. It's a fine art eliminating signal transients, shown at www.commsdesign.com/main/9802fe4 .htm and www.eedesign.com/editorial/1996/pcbdesign9605.html.

Grounding

Make sure that signals travel over a ground path that carries their return signal. All the electrons sent down a signal trace balance the corresponding electrons returning in the ground plane beneath the signal. If the ground plane has gaps in it, then the return electrons must trace a different path back. This creates a loop of electrons moving about, generating more interference. Avoid splitting ground planes in a PCB layout.

- www.devicelink.com/mddi/archive/96/08/011.htm
- www.cae.wisc.edu/~benedict/pcbpres.pdf

Filter the Power Supply

As logic circuitry switches signals from one voltage to another, the power supply struggles to deliver current to each logic node. Since it takes time to move electrons, it makes sense to store electrons in the places likely to need them most. Power supply capacitors are designed to provide this power and to decrease the transients on the PCB. Consult the following URLs to learn how to filter power supplies properly on a PCB.

- www.icst.com/products/pdf//note07.pdf
- www.quicklogic.com/images/pcb_de_1.pdf

Linear Power Supplies

If the design can put up with some inefficiency, consider using linear power supplies. Switching power supplies can generate a considerable amount of interference emitted both as RF and through the power line.

Isolate Noisy Circuitry

Keep very high frequency circuits well away from *input/output* (I/O) circuitry that leads to the outside of the robot. Interference can move right through PCBs to neighboring circuits. Try to isolate I/O circuitry as much as possible from all other sources of interference on the PCB board.

Quiet Motors

Beware of motors with brushes that create sparks. Some motors are more quiet than others. It's a good bet that if sparks are visible when looking from the edge of the motor, it is generating a significant amount of interference. If no qualitative way to evaluate a motor is available, try using the detuned radio method mentioned earlier. A noisy motor will make a radio crackle and pop.

Use Pretested Components

It is possible to buy pretested components, such as power supplies, that have already been tested for emissions. The manufacturers can provide profiles showing the emissions at various frequencies. The testing agencies will often take these profiles into account. If they feel the tests have already been run, they may skip some tests. However, from experience, it seems to be the case that pretested components don't always live up to their reputation. A power supply that has already been tested and certified will actually fail to meet emission specs in a new robot. It's always wise to repeat all the tests from scratch.

SHIELDING

So how do we keep interference inside the robot (and interference from entering)? First of all, we cannot prevent it completely. All packages for electrical systems will allow interference to leak through. The trick is to keep it well below the tolerable levels prescribed by the governmental groups that regulate it. Many techniques exist for limiting the amount of electrical emissions that escape a robot.

Limit Openings in the Package

Interference can leak out of holes in the package, but given the fact that interference waves have a definitive wavelength, there's a trick we can use. Waves cannot easily get through a hole that is too small for them. At most of the frequencies the FCC cares about, holes of up to $1/8$ of an inch (3 mm) in diameter are just fine. If the air holes, for instance, are 3 mm in diameter, they can still provide cooling without letting interference through. If the robot is to be used in environments that permit even less interference, the holes may have to be made smaller again. The higher the regulations go in frequency, the smaller the holes should be. This includes fan holes too; holes are holes!

One point must be made about limiting the size of holes in the package. When we talk about limiting the size of holes to 3 mm, we are speaking about the longest dimension of the hole. If a hole is 3 mm wide and 9 mm long, it will not pass muster. The 9 mm dimension is 3 times too wide. The single worst types of holes in the package are seams. Often, a cover is put on the package and screwed down. The cover may be 30 mm by 30 mm and be fastened down by several screws. Unfortunately, the 30 mm seams will leak like sieves.

The interference that leaks out through such holes can be decreased in a couple of ways. First, it's possible to have a significant metal overlap at the seams. If the package overlaps the cover by more than 1 mm, it's possible to attenuate much of the interference that may leak through. To be safe, have a large overlap. The alternative is to have a spring-loaded metal barrier that acts to seal the seam. Companies sell strips of stamped copper spring material that can be fastened down the length of the seams, much like the weather stripping we use to seal storm doors against the cold wind.

Use Special Connectors

Connectors, and the external wires that will connect to them, are a prime place for interference to escape from the robot. Two characteristics of connectors must be considered:

- First, make sure the connector has a shielded, grounded shell. This means the outside shell of the connector is connected to the chassis and is grounded. The cable that connects to the connector can thus also have a shielded connector and outer metal jacket.
- Second, make sure that all the signals in the connector have attenuating filters in series with them. Don't forget; interference makes no distinction between input signals, output signals, power, or ground. Interference can travel in and out of all types of connector pins. Many connector manufacturers offer versions of connectors with integral ferrite plates that will attenuate high-frequency interference on every pin. The other option is to build filters into the PCB near the connector.

Power Cord Filters

If a power cord or a charging cord is used, consider putting a ferrite filter in series with the wiring inside the robot. Generally, this can be done with a couple of loops of the power wire through a ferrite toroid.

Robots that will be manufactured in quantity will have to be checked for emissions and certified to comply with FCC and CE standards. Several companies can assist with this effort, testing the robot in various configurations. However, the charges for such an effort can run into many thousands of dollars. Without experience, it is unwise to attempt to control such a testing project. Professionals can be hired to represent a novice robot builder during the design and during the testing process. During testing, the testing companies will often overlook obvious fixes that can greatly speed up the testing process. If the design effort has a professional EMI expert on the team during the testing process, the testing technicians can be prodded into action and much money and time will be saved.

Quality Issues

Some years ago, Japanese automakers made major inroads into the American auto market. In large measure, this was due to their persistent attention to quality issues. Every year their automobiles got better and better. Eventually, the American automakers also began to adopt the Japanese quality processes. In recent decades, various names and buzzwords have cropped up, including *Total Quality Management* (TQM), ISO9000, continuous quality improvement, and so on.

Several aspects of quality processes have remained largely constant over all the different incarnations of quality control. Chief among them are the following:

- **Continuous improvement** The process of improving the quality of the robot should not be a one-shot deal. Periodically, the robot's design and manufacturing process should be improved with an eye toward making the final robot better and more reliable. Over time, if everyone on the design and manufacturing team knows that continuous improvement is the goal, all aspects of the robot's reliability and quality will steadily improve.
- **Quality reviews** Once called quality circles, the review process simply schedules periodic examinations of the robot's quality. The team gets together, reviews all reports of problems, and suggests improvements.
- **Empowerment** It is said that everyone on the development and manufacturing team should be empowered to call a halt to design or production if a problem is

suspected. As a practical matter, this may well be giving too much power to stop production lines. But the fact is, unless everyone is on the ball and feels like they can make a difference, then quality processes may not work. Empowerment puts the emphasis on quality first and foremost.

- **Process documentation**   Good quality control systems call for the documentation of the quality process. For a small group, this may prove to be too burdensome. If the design and manufacturing group is 5 to 20 people, then consider adopting formal documentation. Advice and documentation can be found at www.isoeasy.org/ and at www.praxiom.com. A web site discussing some of the basics is located at www.optants.com/tutor/ciptqm.htm.

Testing

Testing is an important aspect of reliability. The word testing has different definitions for every engineer. This is because many kinds of tests exist, and they are all used to accomplish different ends. Many test engineers have been able to make a career out of testing systems. This section outlines the different types of tests.

STRESS TESTS

As we discussed previously, it's possible to stress portions of the robot with various environmental factors like temperature, vibration, and humidity. Many things can be learned from stress tests, including

- **The limits of operations**   At what point will the robot stop working and why? As an example, it's possible to raise and lower the robot's temperature to find out which components will stop working. Further, we can find out whether the components break or just become temporarily inoperative. From such tests, the design can be modified to make the robot more robust.
- **Spec verification**   Will the robot work during a particular stress test? If it does, then it's possible to say with confidence that it will do so again. This is an accepted way to verify that the robot can meet a particular specification. The specification is often published along with the test method.
- **Life testing**   If a significant number of robots are tested, it's possible to develop a statistically valid prediction for their lifetimes. Accelerated baking (at a high temperature) can age components at a fast rate. The components to be baked are slowly taken up to a temperature like 50° C and left to operate for days. Any failure to operate is noted. If enough components are in the oven for a long enough time, it's possible to then develop a predicted failure rate for the component. The

components can be single components or entire robots. The statistics and math behind the statistics are difficult. Some of the techniques are described at www. asnt.org/home.htm.

PERFORMANCE TESTS

Most robots will have specific tasks they should be able to carry out. Some of these tasks will be measured qualitatively and some will be measured quantitatively. Test engineers make a full test regimen and then execute it to determine if the robot passes muster and meets specifications. Each aspect of the robot's performance can be measured, and the quantitative performance statistics can be gathered. Some performance tests are as follows:

- **Full test**   The entire test is executed and may take days to complete. The goal is to get a complete read on how well the robot performs a baseline.
- **Regression test**   This is a subset of the entire test and may be executed many times. The test is short because it must be inexpensive to execute, since it's performed so often. The test is executed every time the robot is changed in any significant way. The goal of the test is to have a reasonable chance of uncovering any errors that were introduced during the changes. Periodically, to gain further assurance, the full test can be executed instead.
- **Unit tests**   Some software engineers segment the software programs into distinct subsections. Each subsection has a specific function, which can be individually tested. Along with writing a function, the software engineers sometimes write a unit test for the function. When the software is compiled, the unit tests are all executed to see if they still work. If the programmers accidentally changed the way the function operates, the unit test for that function will likely fail and alert the engineers of the problem.
- **Use tests**   Designs are not human and can't be confounded by many "what ifs." What if the robot is turned on and somebody forgets to connect a connector? What if buttons are pressed in the wrong sequence? What if the battery wears all the way down? What if the wheels lock up, will the motor burn out?

All these sorts of events should be tried at least once to observe the results. If anything untoward happens, then either the manual should be rewritten to prevent the event or the design should be changed.

Reliability, safety, and compliance are areas where experience counts. When in doubt, seek experienced help and advice. Many technically good designs fail to pass muster when these topics are considered. Plan your approach well in advance.

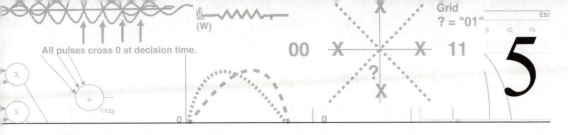

DESIGN STEPS: HLD

Nobody really has a clear, documented record of the thinking that went into the design of a robot, so we'll try to document the series of steps and thoughts in a logical sequence. Others would go about this in a different way, but it makes sense to give a good solid example of how things might be done during the design phase.

Certainly, we should go about writing specifications for the project, just as recommended in Chapter 1. But let's suppose we already have a specification written. It makes sense to take a deep breath at the beginning of the project and just define what success will mean. With that done, and a reading of the specs, it's time to start the *high-level design* (HLD).

Power

Any robot design should start with a thorough analysis of the power requirements. We'll discuss power in a separate chapter, so let's just mention it now in passing. Unless the power source is sufficient to match the requirements, the robot cannot operate properly.

To look at the power, we'll need to look at weight, required activities, locomotion methods, operation time, energy storage schemes, automation, communications, and refueling (recharging).

Locomotion

We'll get into the mechanics of the robot in another chapter, but we should mention it now. We'll need to look at the mechanics needed to move the robot, the power needed to affect movement, the required speeds, and the requirement for reliability. We will need to look at the degrees of freedom required. We can think of degrees of freedom almost like joints in a human limb. The robot will have to bend various directions and must have separate control over each axis.

With the requirements for movement and power estimates in hand, we have all the basics roughed out. We know how heavy the robot is, how much it will have to move, and what sort of power source we will use. This part of the HLD is akin to planning an invasion in wartime, like the invasion of Germany in World War II. General Bradley knew how many tanks were required and how far they had to travel. This allowed him to quickly rough out preliminary plans for the fuel supply.

Automation

Next, with the basic logistics worked out, it's time to look at the automation of the robot. We can assume for the moment that the specifications have already been simplified, so the HLD problems are straightforward. Further, we can assume that computerization is already in the plan. During the HLD, we'll look to simplify things further.

A computer can often take over tasks that might be performed in other ways. If we can move some of the robot's functions into software, we gain two advantages. First, we can delay portions of the design until the software needs to be written. Second, we can reduce costs. Software is free to the extent that software programs can be loaded into robot after robot for free (once we own the software).

Here's a specific example of what can be done in software. Suppose the robot has rechargeable batteries. Further, suppose the specifications call for notifying the operator when the batteries are recharged. This can be handled in two ways:

- We could use intelligent batteries that have the capability to communicate with a computer. These batteries have special connectors that carry serialized signals to a computer interface. We can program the computer to interrogate the batteries and report on their status, but an easier way exists.

- If we are content with the accuracy it affords us, and if we are not worried about the consequences of getting bad information now and then, we can simply simulate the batteries inside the software. We only need to know how long the batteries have been drained by the robot and how long they have been charging. If the simulation errs on the side of keeping the batteries charged, then the computer will be able to perform the function entirely in software. What advantages accrue to the design team? First of all, we will only need standard rechargeable batteries as sold in the store. We will not need special batteries that can communicate. We won't need a battery interface on the computer. The robot will cost less as a result.

Many other functions can be moved from hardware into software. Just be aware that what the computer and the programmers can do is limited. Times will occur when the inclusion of hardware will obviate the need for painful and expensive programming.

Reexamine the robot design during the HLD review process. Have the team meet and discuss the welfare of their new offspring. Bring in outside advisors for the review meeting who may be able to spot things others cannot. Several questions should be addressed, including

- Is it simple enough and reliable? If team members are uneasy about sections of the design, that's a place to start the discussion.

- Take a close look at all the parts that might have high failure rates or might be environmentally sensitive. Reduce the need for those parts if at all possible.

- Reduce the need for risky operations or mechanics. The best mechanical designs tend to be extremely simple.

- Look for places failures could occur. It does not take an expert to sense where a design may have problems.

- Take a look at the requirements for automation. What algorithms will be used?

- Is the software simple enough? Are the programmers running wild? (Oh yes, they will do that given too much sugar!)

- Can the software cause failures all by itself? Software reliability is a major technical arena with conferences, toolsets, specialists, and so on.

- Are there sufficient design margins? Do the actuators, batteries, and computer circuits have more than enough horsepower to achieve their goals? It's wise to over-specify by a significant margin when specifying these items. Most projects expand in a somewhat unplanned way. So get a jump on it and reserve spare resources the robot can draw upon.

Once the basics of the robot have been roughed out, and the HLD has been written down and reviewed, it's time to get fully organized for development. No engineer likes to wait for another engineer's work to be completed, nor do they appreciate being stalled for either decisions or resources. It's important to put together working guidelines and plans that make things work smoothly. Here's one suggested way to help make this happen.

Divide the team up into independent groups. One group could handle the mechanics and power systems. A second group could handle the automation. Have the teams sit down at the beginning and work out all the interactions between the two groups. The following issues should be addressed in this particular case:

- What signals will the mechanics provide to the computer, and what signals will the computer provide to the mechanics?
- Sit down, draw out, and explain all the major movements and functions of the robot in storyboard form. Not everyone will have read the specifications. Further, many people cannot simply read specs and visualize the operations. Some people have to see things and hear them before they will fully understand.
- Discuss which tests will be performed and who will document the test regimens.
- Discuss which *Computer-Assisted Design* (CAD) systems will be used for mechanical and electrical design. Ideally, these systems should be integrated so that it is easier to fit the *printed circuit boards* (PCBs) into the mechanical chassis.
- Discuss how the mechanics will fit inside the robot. Although a CAD system can be used to align things, almost nothing can be a substitute for an audit of the critical areas in the robot. As an example, let's suppose we are designing a PCB that must fit within the robot. Let's further assume that the CAD systems are not integrated, as is often the case. Make a spreadsheet of every interaction point within the robot where the PCB might interact with the mechanics and packaging. By interact, we mean touch or require accommodation. For each of the interaction points, enter all the relevant dimensions for that point into the spreadsheet, including XYZ coordinates. With a thorough tabulation of the interaction points, it is much easier to determine if the PCB will fit within the robot's mechanics without an error. Without such attention to detail, it is very easy to suddenly realize that a post is right where we thought the PCB would go. Make mockups, if need be, out of Styrofoam and cardboard. Just don't let the "customer" see it!

- Agree on the methods of communication between the two teams. Meet as often as necessary to maintain the proper flow of information.

With all these details squared away, a good project manager can keep both teams busy during development. Stay in touch with all the engineers on a daily basis and stay alert. Problems can develop quickly. Move as rapidly as possible toward the execution of the first test regimen and the project should go well.

ENERGY AND POWER SYSTEMS

The power systems of a robot are central to its health, reliability, and effectiveness. The power systems include all the elements of the robot that work together to generate, use, and conserve power. It is very difficult to control the power profile of a robot. It's critical that the design team starts early on a plan for power control. Further, it's critical that one engineer be in charge of the effort. Ideally, if a computer is involved, give the task to the engineer that can control the power-saving features of the computer processor itself. Every component of the robot, down to the last nut and bolt, affects the power consumption. We'll get into just why that is the case later.

For the moment, let's take a big step back and perform a mental exercise. Imagine the robot we want to build. Visualize its form, shape, and mass. Now let's take off our collective eyeglasses and view the robot as a big, fuzzy hunk of metal and plastic. It's just a mass of material, portions of which may move from place to place. Will it have enough fuel to get where it's going and to perform its task? With a battery-powered robot, this is a critical question. Viewing the robot as a single mass makes it easier to make sense out of the preliminary energy calculations.

Energy

When the early rocket scientists first began to build rockets, they were immediately confronted with some very basic laws of physics. How, for instance, could they put a satellite into orbit? How could they put two astronauts on the moon and get them back? Eventually, it all boiled down to one consideration: energy. Auditing the energy within the robot is a great way to approach the design of its power systems.

The energy the scientists had to start with was rocket fuel. The Apollo moon-landing problem was to take two astronauts and the *Lunar Excursion Module* (LEM) up to the moon. How much fuel would be needed and how would it be done? They probably sat down with a single pad of paper over lunch and roughed it out in 10 minutes. Lunch probably went something like this.

They figured out the weight of the LEM and the astronauts at around 48,000 kg. From that weight, they could figure out how much fuel it would take to move the LEM from earth orbit up to the moon. Further, they could estimate the energy required to lift the LEM and the Apollo spacecraft (129,000 kg) up into earth orbit in the first place. They needed an efficient way to accomplish the task and came up with the three-stage Saturn rocket concept. Shedding the Saturn rocket stages on the way up into orbit obviated the need to carry the entire rocket's weight into orbit. I'm sure they finished the raw energy calculations in just a few minutes. They came up with the requirement for a three-stage Saturn rocket and crawler standing 111 meters tall and weighing 6 million kilograms (about 6,000 tons). Then, I'm sure, they sat back and ordered another round of margaritas!

The point is, the calculations are not hard, and they don't take too long. We should be able to rough out the energy requirements of the robot rather quickly. But where do we start?

The very first thing to be done, much like the rocket project had to do, is to forecast the energy that will be required. We know the approximate size of the robot we want to build. We also know roughly what sort of motions and actions the robot will have to take. We can forecast the amount of energy the robot will use for movement once it's designed in two different ways: using calculations or using empirical measurements.

CALCULATIONS

By looking at the mass of the robot and knowing the actions the robot must take, we can often calculate the energy that will be required. For example, if we know the robot weighs 50 kg (batteries included) and must climb a 6 meter ladder 10 times a day, we can determine the potential energy required to climb the ladder using the formula PE = m × g × h:

$$PE = 50 \ kg \times 9.8 \ m/s^2 \times 6 \ m = 2,940 \ joules$$

Since 1 joule equals .000278 watt-hours, 2940 joules equals 0.817 watt-hours. Table 6-1 outlines the watt-hour ratings for rechargeable batteries.

Certainly, many other battery technologies are available, but the preliminary calculations show that just one AA NMH battery should carry enough energy to take a robot up the ladder once. We require 0.817 watt-hours of energy and the battery can contain 1.8 watt hours. We have a margin of about 2 to 1. That's not too bad, hauling a robot the size of a 12-year-old boy up a long ladder with one battery. Clearly, to do it 10 times, we'll need 10 × 0.817 watt-hours, or 8+ watt-hours. So we'll need a couple of D-size NMH batteries to provide 15+ watt-hours. We'll see in a bit that a margin of 2 to 1 may not be enough, however.

The astute observer would note that adding more batteries to the robot alters the weight! That's quite true. Simply add the battery weight to the robot's weight, and perform the calculations again. Eventually, everything will pencil out.

TABLE 6-1   Rechargeable batteries' watt-hours

BATTERY SIZE	WATT-HOURS	BATTERY MATERIALS
D	4.0	Nicad
D	7.8	NMH
AA	0.6	Nicad
AA	1.8	NMH

EMPIRICAL MEASUREMENTS

One other way to estimate the power the robot will need is to literally build a model of the robot and try it out. Practically speaking, we do not have to build the entire robot; rather we can simulate it with a hastily built mockup. It would suffice to just build the drive mechanisms and load down the simulated chassis with the proper amount of weight (perhaps with bricks). Then the simulated robot can be put through its paces and the power drain can be measured directly. This will prove to be quite an accurate way of gauging the amount of energy that will be required. It takes into account almost all the inefficiencies that can throw off an energy prediction that might be only calculated.

COMPARISONS

Sometimes we can find systems that must perform tasks similar to what our robot must perform. For instance, if the robot must weigh 3,000 lbs. and carry 4 people up a mountain road, we can just look to a similar sized car and try to emulate its engine and mileage. If the robot must shred celery into small edible bites, we can take apart our Cuisinart and see what kind of motor it has. For that matter, if the comparisons are very close, perhaps we can chop down a Volkswagen or a Cuisinart, build it into our robot, and be done with it!

Don't forget that we have been calculating and measuring the energy required to move the robot. We must also provide energy to power the computer systems, sensors, taillights, and all the other circuits on the robot.

Once we have an estimate of the energy that's required, we must back off a bit and add some design margin to the robot. As a practical matter, theoretical calculations of work are very rough. Motor inefficiencies, friction, and many other inefficiencies use up energy from the battery in useless work. It makes more sense to have a 4:1 (or higher) ratio of energy to required energy. Translated to efficiency, we only expect our robot to be 25 percent efficient. If the robot is going into space, the designers will want to do better. If the robot is going across the room, more margin for error exists since it can be serviced or redesigned. Figure 6-1 shows a typical 20-watt DC servo motor operating at about 25 to 50 percent efficiency. Note that efficiency depends upon the torque that the motor must exert. Also, peak efficiency does not occur at maximum mechanical power output.

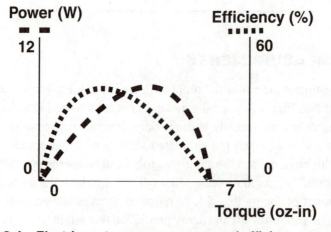

FIGURE 6-1 Electric motor curves: power and efficiency versus torque

Energy Sources

Energy can be acquired and stored in many ways, but we won't go into the different types of battery technologies here and now. Articles about batteries and alternate power sources can be found at the following web sites:

- www.powerstream.com/BatteryFAQ.html
- www.powerstream.com/tech.html
- www.motionnet.com/cgi-bin/search.exe?a=cat&no=1308
- www.batterystuff.com/battery/battery_tutorial.htm

Instead of talking about power sources directly, let's list the characteristics we should pay attention to in the search for power:

- **Weight versus energy**   The weight of the power source is a prime concern in satellites, mobile systems, and portable systems like laptops. Some battery and fuel cell systems will be lighter per watt-hour than others. Certainly, any mobile robot should be as light as possible to avoid expending unnecessary energy.
- **Capacity**   How many watt-hours can the battery store? How is the end of its useful life measured?
- **Peak currents**   Some batteries are better than others at delivering large peak currents. Besides checking the magnitude of the peak current, determine how long the battery can sustain such a current. It may not be able to do so for very long.
- **Lifetime**   What mechanisms may cause the battery to fail as it ages?
- **Temperature**   Will the battery function at sufficient levels over the required temperature range?
- **Recharging**   How is it recharged? Are there any special requirements?
- **Cost**   How expensive is the battery and can it be readily replaced?
- **Safety**   We discussed before the many hazards that batteries can present. Have the proper safeguards been taken?
- **Warm-up**   Will the battery require any warm-up time to function properly?
- **Metering**   Is the battery smart enough to communicate with the computer? Failing that, is the battery relatively predictable in its charge/discharge characteristics? We may have to simulate the state of the battery in the robot's software.
- **Availability**   How special is the battery? Will it be supported by the industry for some years to come? Will replacements be available on the open market?

Like humans, robots will only work well when fed enough and exercised within their capabilities. Understanding energy, power and motion are key to building a successful robot.

7

ENERGY CONTROL AND SOFTWARE

Considerations

Most robot control system designers make an attempt to minimize power consumption in the robot. This is true whether the robot is battery powered or not. If it is battery powered, then energy control is generally a critical part of the software design that must be held in the forefront of all software design considerations. If software is written at first without regard to energy control issues, it generally will need a rewrite. Even mundane database and housekeeping software needs to be written with energy control in mind.

Crafting an energy control strategy is a fine art. It can be very difficult to scale back the energy consumption of a robot and its computer. The finest designs use no extra energy than is absolutely necessary. To accomplish this feat cleanly, project managers and design engineers must pay attention to a few rules of thumb.

PRIORITIES

Put energy control first. To successfully conform to energy control requirements, it is almost always necessary to put this issue in the forefront. If energy control ever becomes an afterthought, it just won't get done right at all. Many decisions are made along the way that could preclude retrofitting a control system with energy control at a later date.

LEADERSHIP

Keep one experienced person (the "energy czar") in charge. As we mentioned in a different chapter, the project must have one single person in charge of energy management. Ideally, it should be the person who best understands the energy control capabilities built into the processor chip. The key word here is *experienced*. Although I deplore the tendency of firms to exclude good engineers who just don't have direct experience in the technology of interest, this case does require such an approach. Energy control in a portable computer system (like in a robot) is a very complex task, one requiring an experienced hand. If multiple engineers are involved, they must also be coordinated by this one experienced person.

PLANNING

Energy control will only succeed if the specifications are crafted with its specific goals and requirements in mind from the very start. The energy czar should be in on all the early architectural, specification, and planning meetings.

BE CONSERVATIVE

Don't underestimate the effort. The czar will have software to write and tests to perform all the way through the project. It's a risky portion of the project and difficult to finish. It's also not unusual for difficulties to crop up late in the project. Even perfectly working software may suddenly fail (increasing the energy draw) for unapparent reasons. Have patience and expect to work hard on this portion of the project. Test and retest the energy draw with each new engineering change.

TECHNOLOGY SELECTION

Go with existing processor power saving technology. Complete control of energy in a computer system generally requires the proper choice of processor. Some processors are better than others for this task. The point is, if the processor is designed to enable the energy control you require, then it probably has special-purpose hardware built in. Often, software drivers for the energy control hardware are already available. The over-all energy control algorithm must take advantage of these processor features. Attempting to circumvent them or to use them in nonstandard ways will likely mean ruination. The processor designers probably had a difficult time getting things right and it is easy to break their design. Stick to the basics and don't be afraid to call the proces-sor company.

CENTRALIZATION

Try to centralize the energy code. To function properly, energy control software must often be spread over most of the software in the robot's computer. Portions of the code are in the flash memory boot code, the operating system, and the application code. As the code spreads out like this, it starts to get "holes" in it. It becomes more fragile in terms of its capability to provide proper energy control. Simultaneously, if multiple pro-grammers are involved, programming discipline becomes more difficult. We must pro-vide a simple, understandable *application programmer's interface* (API) that will attend to energy control properly. Fortunately, a couple of remedies are available:

- **Keep it tight**   If the API code is written from the ground up, try to keep the API small and confined to the lowest levels of code (the *Basic Input/Output System* [BIOS] and a few drivers).
- **Keep it simple**   If you must build your own processor drivers, and can get away with it, don't try to implement every feature the processor offers. Get the most aggressive power-saving features of the processor working well first. If the other features are also desirable, add them later if you can.

Now we should look at strategies for power control. Where do we start?

It makes sense to start with the specifications. While the specifications are being written, we will probably have a sense of how difficult the energy control problem will be. It makes sense, right then, to look at energy control in a larger sense and decide on the overall approach. The government even has specific programs to encourage the industry to decrease energy consumption (see www.energystar.gov).

Semiconductor companies and operating system companies work together to provide a coordinated approach to power control. As an example, the community defining the PC architecture has done this over the years. Committees meet once a year to ratify the changes that all parties will make in their designs and publish documents specifying the changes. The Windows Hardware Engineering Conference (WiNHEC) is one of these committees (www.microsoft.com/winhec/default.mspx).

The following URLs describe industry initiatives that are designed to save power in computer systems.

- Mobile technology www.intel.com/pressroom/archive/releases/20020304comp.htm
- Instant on www.intel.com/technology/IAPC/index.htm

It is a very complex task to bring the power consumption of a large computer system way down. It's much easier if the computer system is designed specifically for low power. The system components probably were developed specifically for an energy-saving situation. All the new software can be written to suit the requirements.

In PCs, the impetus for the setting of standards came from the portable PC market. The trouble was that the entire architecture evolved from a desktop PC architecture that had few power-saving features. As such, the design is too leaky to be useful as an embedded design.

Energy Conservation

It makes sense to model the overall type of energy conservation states that the robot will have. Computer companies give fancy names to these states, so I thought I would make up my own names too. The energy control systems of computer control systems and robots can be described using the same terms. Each energy state can be described with a set of characteristics including how it uses and conserves energy, how the state is used, and how the programmers interact with the software.

ENERGY PIG

This would be a robot that always runs at full throttle, whether it needs to or not.

- **Energy use**   All the lights are on and no attempt is made to conserve energy to speak of. The computer runs full tilt all the time, and the motors are continuously trying to servo into the right position. Most of the components are using energy at their maximum consumption rate.
- **Conservation tactics**   No intentional energy-saving methods are designed in.
- **Method of powering up**   None exists; it's always awake and active.
- **Delays**   The robot is always powered up, so no processing delays take place.
- **Special uses**   Since the software always runs, no restrictions are made on the types of systems that can be created this way.

- **Software interaction**   The software is unaware of power-saving measures, so no energy API is needed.

ENERGY AWARE

This is a robot that incorporates one or two of the easier energy-saving tricks. Perhaps the motors shut down if it's not moving. Not much time was spent designing the energy control.

- **Energy use**   A few conscious attempts at energy savings are made.
- **Conservation tactics**   The computer might use one lower-power state that enables it to conserve power during its idle loops. The software turns some major components off when they are not in use.
- **Method of powering up**   Generally, both interrupts and application software can bring the robot quickly back to a full operating state.
- **Delays**   The robot powers up very rapidly since all the major software environments remain loaded in memory. Processing delays are very short.
- **Special uses**   The software has few restrictions, so most robot applications can use this type of energy control if it meets their requirements for conservation.
- **Software interaction**   The software has minor API hooks to enable applications to wake it up and put it to sleep. Since the software environments are always resident, few other API calls are needed.

ENERGY EFFICIENT

This is a robot designed from the ground up to be efficient in its use of energy. The motors rarely are powered up, the computer only runs when it has to, and all subsystems are designed to use very little power.

- **Energy use**   Energy use is greatly minimized. Most energy minimization techniques have been used; only the most difficult ones are left undone.
- **Conservation tactics**   The computer makes use of the most aggressive energy-saving states it can. The software turns all major components off when they are not in use.
- **Method of powering up**   Generally, only interrupts from the timer or outside stimuli will bring the robot back to a full operating state.
- **Delays**   The robot takes a while to power up because many of the major software environments must be reloaded into memory. Processing delays are not short.
- **Special uses**   The software control algorithms must be able to withstand significant delays because of the sleep state of the processor.

- **Software interaction**   The software has a significant number of API hooks to enable applications to store and retrieve environments, wake up the processor, and put it to sleep. The API generally has hooks for both the boot *Read-Only Memory* (ROM) and the application software.

ENERGY MISER

This is a robot that uses the bare minimum amount of energy to accomplish its tasks. The best examples of this are space-traveling robots, optimized for solar power and the conservation of energy. Mobile robots often have a similar design, especially if they operate unattended.

- **Energy use**   Energy is completely minimized.
- **Conservation tactics**   Every subsystem is turned off when not in use. The batteries have very long lifetimes and the robot will not drain them if left immobile. The computer makes use of the most aggressive energy-saving states that it can.
- **Method of powering up**   Generally, only interrupts from the timer or outside stimuli will bring the robot back to a full operating state.
- **Delays**   The robot takes quite a while to power up because all the major software environments must be reloaded into memory. Processing delays may be significant.
- **Special uses**   Generally, the application must be able to tolerate significant delays on the part of the computer control system. This tends to restrict the use of such drastic energy-saving methods to specialized situations such as handheld systems and unattended robots.
- **Software interaction**   The software has a significant number of API hooks that enable one or more deep-sleep energy states. Most of the application code must be specially written in light of it.

Hardware Considerations

Nothing but hardware can use energy. Software cannot use energy directly without commanding hardware to do to. The hardware itself must be well constructed so as to not waste energy.

POWER SUPPLY

Certainly, the power supply is a very important place to start when trying to save energy. In fact, any heatsink in the robot is a prime place to try to save energy. Energy wasted in heatsinks creates two problems. First, it's a direct waste of energy. Second, the heat must be dissipated. In battery-powered robots, these considerations are especially important.

Regulation

A couple of different types of power supply exist, but the bottom line is that they all waste some energy in an attempt to regulate the voltage going into the robot's control circuitry. The robot may be powered from power lines or from batteries. In either case, the motors and control systems generate electrical noise as they turn on and off. Motors, in particular, cause large power surges when they turn on and off. We all have seen these sorts of surges when a large appliance starts up and the lights flicker temporarily. The point is, even battery voltages may vary unacceptably during the operation of a robot. We may not be able to feed the battery voltage directly into sensitive control circuits. We also know that power from the power lines is not well behaved either; it needs cleaning up. For all of these reasons, robots generally have some sort of voltage regulation circuitry or power supply. That said, before we go into a discussion of what type of power supply to put in, let's explore *not* putting a power supply in. What are the techniques we can use to avoid a power supply?

Battery Power

Batteries are fairly stable voltage sources all by themselves. They do, however, vary under certain conditions:

- **Charge level**   The voltage on batteries will decrease as the level of energy in the batteries decreases. Different battery technologies will decrease at different rates over time. They have characteristic discharge curves with a voltage that decreases in a predictable manner as the charge is drained out of them. Most rechargeable batteries decrease rapidly during the first 5 percent of the discharge curve, level out for 85 to 90 percent of the curve, and decrease rapidly in the last 10 to 15 percent of the curve. Figure 7-1 displays a representative battery discharge curve.
- **Voltage levels**   A few points need to be made about voltage levels. First, if the battery is being charged while it's inside the robot, great care must be taken. An inexpensive battery charger, charging an open battery, doesn't limit voltage. Instead, it would attempt to deliver a much higher voltage, which spreads throughout the robot. Batteries act like voltage limiters when they are being charged up and will limit the voltage of a cheap charger. But if the battery opens up (or is simply removed), the charger may fry the rest of the robot with abnormally high voltage!

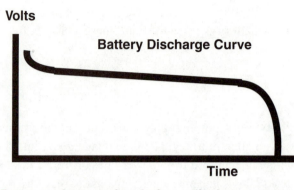

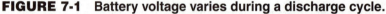

FIGURE 7-1 Battery voltage varies during a discharge cycle.

This is especially true if no power supply circuitry exists and the robot is running off the battery directly.

■ **Internal resistance**   Batteries will all have different internal resistances. This behaves much like a resistor in series with the battery. As the battery ages, this resistance may change. When a motor or other heavy load, places a sudden demand on the battery for current, the voltage of the battery will change quickly. Make sure the rest of the robot's control circuitry and sensitive instrumentation can take the sudden voltage transient on the power supply.

■ **Lifetime**   Don't forget that the ability of batteries to store energy will change over time. Many types of batteries (with different internal chemistry) will lose their capability to store power as the battery ages. Within the battery, chemicals, gases, and metals migrate or slowly corrode so they are no longer able to fully contribute to energy storage. Make sure the robot's circuitry will be able to function just as well when the robot and its batteries reach old age (see Figure 7-2).

Power Requirements

If we are trying to power a robot using just the battery as our power supply, we need to limit the number of different voltages that will be needed within the robot. This may mean that all the electrical components must be selected so they can work off the same voltage. This becomes quite a challenge when we try to pick motors, sensors, and computers that all have similar requirements for voltage. So what voltage should we try for? High voltage, for example, is not a good choice for running computers or most sensors.

Motors   To complicate things further, low voltage does not work well to move motors. We can use very low voltage drop *Field Effect Transistors* (FETs) to control the motor

FIGURE 7-2   An elderly robot toy

puters that all have similar requirements for voltage. So what voltage should we try for? High voltage, for example, is not a good choice for running computers or most sensors.

Motors   To complicate things further, low voltage does not work well to move motors. We can use very low voltage drop *Field Effect Transistors* (FETs) to control the motor windings and keep the efficiency up. Semiconductor companies sell chips specifically designed to control motors in an efficient manner. Some low-voltage motors are available, but it would restrict our choices. Many motors are available that require a 12-volt drive. Fewer are available that will work with a 5-volt drive. The requirement to turn the voltage off and on to a motor further complicates the supply question because most voltage switches (and wiring) will also drop the voltage available to the motor.

Several alternatives also exist to traditional motor technology. Esoteric motors may be fun to investigate, but use them with care. The motor found at www.drives.co.uk/news/prodnews/news_prodnews148.htm, for example, uses piezoelectric power to create movement and uses low voltage.

low-power technology as discussed previously, also has two other nice characteristics we could use:

■ **High noise margins** Most semiconductor technologies, such as bipolar and some FET technologies, have strict requirements for the voltage levels that can be present on the circuit board. Processors and integrated circuits made from these technologies have signals that only vary over a small percentage of the power supply voltage range. If the signals vary outside of these ranges, then logic errors might occur and the robot will malfunction.

CMOS FET technology can tolerate a much wider range of signal voltages. With some restrictions, CMOS logic gates regard signals above 50 percent of the power supply voltage as logic one, and signals under 50 percent of the power supply voltage as logic zero. The power supply can even change voltage (within bounds) and CMOS logic gates will still work just fine. It may be difficult to find off-the-shelf computers built with just CMOS because the competing bipolar technologies have the bulk of the commercial market, but certain manufacturers concentrate on CMOS and other logic families cater to the requirements of portable and robot applications. It should also be noted that some logic families work better than others in the presence of nuclear radiation. If your robot will be going to truly hot locations, give CMOS a good look!

Here are some PDF files that discuss high noise margin logic:

■ www.ece.pdx.edu/~greenwd/AN_375.pdf
■ http://lorien.die.upm.es/~macias/docencia/datasheets/info-familias/
  hc-cmos-dc-characteristics.pdf

■ **Power supply range** CMOS technology will work over a relatively wide range of power supply voltages. Most *single board computers* (SBCs) work off 5 volts, but it's not impossible to find boards that will accept a wider voltage range. Automotive designers have been using CMOS and related chip technologies for years, even though they are not stuck for energy.

Here are some sites providing information about CMOS logic families:

■ www.bychoice.com/cmos.htm
■ http://us.st.com/stonline/prodpres/standard/stanlogi/hcmos.htm
■ www.electronicstalk.com/news/sra/sra100.html

Power Regulation

Energy is not always available in a form that can be used successfully. Often, it has to be transformed and tamed. This can be done in a few different ways and some are more suitable than others for battery-powered robots. It's time to review power regulation and, in the bargain, we can take note of regulation techniques that are good for robots.

The central problem to be solved in power regulation is to prepare an untamed energy source to provide tame power for the robot. Certainly, the type of tame power needed by the robot will vary. In some instances, the robot's circuitry can use unregulated *Direct Current* (DC) or *Alternating Current* (AC) to power components like motors or solenoids. But in most cases, the robot's components will need well-regulated DC power. This type of power is generally specified by the voltage, the acceptable voltage range, the current available, and the level of ripple that can be tolerated. For some 5-volt DC supplies, the specifications might read: "5V +- 0.25V, 5A, 25 mv pp ripple." This is a power supply that can deliver 5 amps into the robot at a voltage between 4.75 and 5.25 volts with only 25 millivolts of ripple noise. The ripple noise is often 60 Hz of noise (on supplies driven by the AC power) or a higher frequency from a switching action that will be discussed shortly.

Unstated specifications for a power regulator include the following:

- **Efficiency**   Although our example power supply might deliver 25 watts into the robot (5V × 5A), it might require a feeder wattage of 40 watts to do so. That would make its efficiency 25/40 = 62.5 percent. The power regulator alone wastes 37.5 percent of the energy.
- **Emissions**   Power supplies generate interference (electrical noise and radiation), which propagates out all the power connections and through the air. Since compliance with regulatory bodies is often required (as mentioned in Chapter 4), we must pay attention to the power supply as part of this effort.

Types of Regulators   Power supply regulators are available in many forms, including the following:

- **Linear regulators**   Linear regulators are an older technology that is well characterized. One or more large transistors take the unregulated power at a higher voltage (Vin) in one side and deliver regulated power at a lower voltage (Vout) out the other side. By and large, since the current flows linearly through the power supply, Efficiency = Vout/Vin.
  Generally, the larger the difference between Vout and Vin, the better the power supply, keeping noise spikes on Vin from getting to Vout. Unfortunately, this lowers the efficiency. Also, more cooling may be necessary; the power supply transistors may need larger heatsinks. Linear regulators are relatively simple and can be reduced to a single three-terminal component with connections for Vin, Vout, and Ground. They do not generate significant electrical interference, but they are not very efficient as a rule.

The following PDF files contain basic information about both linear and switching power supplies:

- www.web-ee.com/primers/files/f4.pdf
- www.web-ee.com/primers/files/AN-556.pdf

An offshoot of linear regulators is the *Low Drop Out* (LDO) regulator. LDOs are linear regulators that expect a very low difference between Vin and Vout. They are used primarily for the local regulation of voltage or in situations where Vin is very low and Vout must be as high as possible. Since the efficiency is high (Vout/Vin), LDOs generally do not need large heatsinks.

LDOs can be used for distributed regulation. Instead of having a single power supply in the robot, Vin is distributed throughout the robot and sent to several LDOs, which provide regulated Vout power to different parts of the robot.

- **Switching regulators**   Switching regulators are generally more efficient than linear regulators. In addition, they can perform feats like making Vout higher than Vin, but this does not mean the efficiency is higher than 100 percent. Because current does not flow linearly through a switcher, the efficiency cannot be easily computed.

Switchers basically take Vin and convert it to a high-frequency AC voltage waveform. This high-frequency current is transformed in various ways to a raw DC voltage that can be higher or lower than Vin. Then the AC components are removed to form Vout, an action made easier because these AC components are high frequency and are easily filtered out. The following PDF files have further explanations of this process:

- www.web-ee.com/primers/files/webex9.pdf
- www.web-ee.com/primers/files/f5.pdf

Switchers can run at a very high efficiency (above 90 percent) when used carefully. In practice, don't count on achieving the claims made by the manufacturer. Count on 75 percent and be surprised if the real number comes out higher. But in a robot, this type of power supply can conserve energy. The downside is that switchers will generate significant amounts of electrical interference of all types.

PROCESSOR

All further considerations of hardware energy savings must start with the processor. The processor has several energy-saving features, which we have discussed before, and they are outlined in the following sections.

Power Supply Voltage

Most low-power processor chips designed for energy-efficient systems can function with very low voltages. We'll see why this is important when we discuss CMOS logic. Suffice it to say that energy consumption is proportional to V^2. This square law of physics pays us great dividends as we move to lower and lower voltage systems. If we can cut the voltage in half, the energy consumption goes down by a factor of 4!

Varying Voltage

Further, some processors can function while the power supply voltage varies. If the processor has this feature, we can take advantage of it in the following way. Because higher voltages can charge up capacitances in the logic chips faster, the computer can run faster at higher voltages. If the computer has little to do, we can lower the voltage and decrease the clock frequency, and the energy draw goes way down. As long as the processor can get its work done in the allotted time, then the robot will function properly and all is well. In the mean time, a great deal of energy will be saved.

To take advantage of this feature, the power supply must be under software control. It must initialize to a suitable voltage and then provide the proper controls that will enable the computer software to alter the processor power supply voltage to acceptable levels. It's possible to get by with a single digital input that alters the power supply voltage. Just make sure that the slew rate of the power supply voltage (the first derivative) is small enough and remains within the limits the processor can accept.

Varying Clock

Processors can be built out of CMOS. All logic families have a basic building block called an inverter. The CMOS inverter is special in that it does not enable the current to flow except when it changes state. Thus, if a CMOS inverter stays static as logic one or logic zero, it will not use energy. However, when it changes state, the capacitance within the inverter must be charged up (changing to a 1) or discharged (changing to a 0). When this happens, a distinct amount of energy is used up in the capacitance of the inverter. The energy in this capacitance is

$$Ecap = 0.5 \times C \times V^2$$

where V is the power supply voltage and C is the capacitance of the CMOS logic inverter gate.

Since this amount of energy is dumped every time the CMOS inverter changes state, the power exerted is proportional to

$$P = Ecap \times f = 0.5 \times C \times V^2 \times f$$

where f is the frequency of the processor clock.

Since the capacitance C is fixed by the CMOS process, it's clear that our best hope for power savings is to decrease V and f. Some modern processors are built to withstand this. Changing their power supply voltage and changing their clock frequency will decrease their power consumption.

Care must be taken, however, that the processor clock is not used for any fixed frequency processes within the robot. Communication interfaces, for example, often require a special fixed frequency for operations. Make sure these interfaces have their own fixed clock frequency. The central clock of the system can first feed into these communication interfaces and then it can be divided down for the processor. Some processors have all this clock division circuitry internal to the processor.

The voltage and the clock can be ramped up and down to fit the workload of the processor. Figure 7-3 shows the method of ramping voltage or the clock up and down, and the relative effect on the processor performance. The same amount of work gets done in the second graph, but since the voltage is half, the power dissipation for that

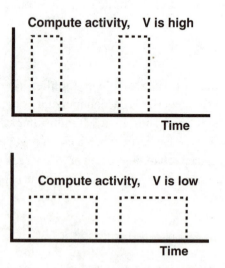

FIGURE 7-3 A computer can save energy by running longer at a lower voltage.

work is a quarter of what it would have been. An interactive tutorial on CMOS can be found at http://tech-www.informatik.uni-hamburg.de/applets/cmos/cmosdemo.html.

Processor Power States

As we've mentioned before, processors have distinct energy-saving states. Pick a processor that has the types of energy-saving states that will match the operating requirements of the robot you are building.

Further, if the robot is to operate in a remote environment (with some remote control), consider giving the operators the ability to alter the processor power state from afar. This is a feature that may just "save the bacon" of the project team someday. Those reading translations of this text are now rather quite confused at this porcine idiom. As an experiment, I have fed this paragraph to a computerized translator and translated it from English to German and back. Before considering building a robot translator, read the next paragraph. I'm sure it will give my editor fits. The two translations follow. Try it yourself at http://babel.altavista.com/tr.

Here's the automatic translation into German:

Weiter wenn der Roboter in einer Aussenstelle funktionieren soll (mit etwas Fernbedienung), betrachten Sie, den Operatoren die Fähigkeit zu geben, um den Prozessorenergie Zustand von fern zu ändern. Dieses ist eine Eigenschaft, die gerade save der Speck kann. von der Projektmannschaft eines Tages. Die Leseübersetzungen dieses Textes sind jetzt an diesem schweineartigen Idiom eher ziemlich konfus. Als Experiment habe ich diesen Punkt zu einem automatisierten Übersetzer eingezogen und ihm von englischem zum Deutschen und zur Rückseite übersetzt. Bevor Sie einen Roboterübersetzer errichtend betrachten, lesen Sie den folgenden Punkt. Sicheres I.m gibt er meinem Herausgeber Sitze. Die Übersetzung folgt. Versuchen Sie sie sich an http://babel.altavista.com/tr.

Here's the automatic translation back to English again:

Far one if the robot in a branch office to function is (with something remote maintenance), regards you, to give to the operators the ability in order to change the processor energy condition from a distance. This is save a characteristic, the straight the bacon can from the project crew of a daily. The read translations of this text are now considerable at this pig-like Idiom rather konfus. As experiment I drew in this point to an automated translator and him from English to the German and to the back translated. Before you regard a robot translator establishing, read the following point. Safe I'm it gives to my publisher of seats. The translation follows. Try it on http://babel.altavista.com/tr.

Oh well.

Processor Power Draw

Some processors are simply designed for low power operation, and others are not. Don't even bother considering processors that do not have the types of features necessary for low power operation. Restrict the search for a processor to suitable energy-saving processors.

Memory Types

When selecting memory technology, pay attention to the power draw of the memory chips themselves. In particular, some flash memory chips have a built-in energy-saving feature. They will move to a low power state if they are not accessed within a certain time period. This can significantly decrease power consumption with little effect on the operating speed of the processor.

SUBSYSTEM POWER CONTROL

The robot's subsystems should be designed with integral power control switches. The processor, under software control, should be able to turn off the power to unused portions of the robot. If, for instance, the robot will be still for a while, we may be able to turn off all power to the actuators and motors. If the robot does not have to sense anything for a while, we can turn off the sensors. A variant of this sort of power control switch is a "dead man" power controller that will turn off subsystem power unless the processor commands otherwise. This is useful in situations where the processor may bomb or if the application software simply forgets to do the proper housekeeping. Remote, unattended robots need this sort of hardware feature on subsystem power control to avoid accidentally draining the batteries.

DRAIN ON INTERROGATION

Try to use sensors that do not consume power unless they are being interrogated. For simple digital inputs, consider using tri-stated processor inputs. Often, it's possible to avoid any energy drain except during the brief period where the processor is interrogating the input.

PATH CHECKING

During the design of the robot, be sure to check every single path that current might take to ground. Often, sneak paths can unexpectedly develop that can drain a battery. Don't assume that all wires, connections, and components are one-way conductors of current. Often, current will flow backwards through a component to provide an unforeseen path. This can drain a battery completely. Robots designed for remote locations (like Mars) are routinely examined for these sorts of sneak paths. For critical missions, determine what happens if a component fails completely. Will it fail as a short? Will it fail open? What is the backup plan for preventing energy losses in such an occurrence?

SENSOR THROTTLING

Since the computer cannot pay constant attention to the sensors anyway, consider turning them off when not in use. Be careful though; choose sensors that have no warm-up time. Often, sensors will drift for a while after they are turned on. If the sensors have integral, internal references and remain accurate with power cycling, they may work well. If the computer must recalibrate the sensors every time they are turned on, it may not be worth it.

PERIPHERAL POWER CONTROL

Many peripherals are available with internal power control circuitry. Sometimes the power controls work automatically within the peripheral, and sometimes the program controls them directly. Such peripherals are as follows:

- **Hard drives** Hard drives can be turned off so they spin down. Most computers offer this option now. Once the disk spins down, it will take a few seconds of latency time for any new data; the disk must spin up to speed before data will be available.
- **Displays** Most computers now have control over the display's consumption of power. On laptops, the backlighting is controlled and desktops control the monitor itself. These components use up quite a bit of energy. If the robot has a requirement for a display, make sure the relevant controls allow control of the energy consumption.
- **Communication interfaces** Communication interfaces carry data into and out of the robot. For robots that are short on available energy, the communication interfaces must be thought through very carefully. One of the most difficult problems to work through is monitoring the communication inputs. It takes energy to monitor a communication input continuously. The next section covers a few developments that may help with this problem.

- **Displays** Most computers now have control over the display's consumption of power. On laptops, the backlighting is controlled and desktops control the monitor itself. These components use up quite a bit of energy. If the robot has a requirement for a display, make sure the relevant controls allow control of the energy consumption.
- **Communication interfaces** Communication interfaces carry data into and out of the robot. For robots that are short on available energy, the communication interfaces must be thought through very carefully. One of the most difficult problems to work through is monitoring the communication inputs. It takes energy to monitor a communication input continuously. The next section covers a few developments that may help with this problem.

Spy-hopping

Some whales have an unusual practice of rising out of the water to see what is going on above the water line (see Figure 7-4). It gives them some visibility they might not have while underwater. Interfaces also spy-hop to detect network activity. The interface only looks at the network periodically so energy use is minimized when no activity exists.

FIGURE 7-4 A whale spy-hopping

need bother watching its receiver until the appointed time. Such a communication protocol can save quite a bit of energy. Certainly, both ends of the link must have accurate, free-running clocks to remain coordinated. An alternative is to use a commonly available clock such as one transmitted by GPS satellites that is available all over the world.

SOME NOTES ABOUT SPY-HOPPING

Spy-hopping is basically a way for the robot to periodically sample the world in which it must function. As we will see in Chapter 8, sampling can easily get the robot in trouble. Two conditions make it possible to use this technique. First, conditions must be well known for sampling to be effective. Second, the control system must be able to function properly with the limited amount of information that proper sampling techniques afford.

We should note at this point that spy-hopping is inherently a type of polling. The robot's control system takes on the responsibility of watching events and catching them as they happen. The processor software goes to each interface periodically and "polls" it to determine if it needs attention. This control method is distinctly different from interrupt-driven control systems where it is up to the event itself to notify the control system that action is needed. Interrupt systems also are capable of low-power operation.

Since spy-hopping relies on sampling, an inherent response time delay is built into the control system. If an event of interest occurs, it will be some time before the processor wakes up and polls the sensors monitoring the event. As long as the event lasts long enough to be detected, the processor will catch it and act properly. A delay will take place, however, which might be as long as the spy-hopping interval. As long as the control system can perform its tasks effectively in the face of the delays, no problems occur.

ADAPTIVE SPY-HOP DUTY CYCLE

The robot's control software can adapt to a changing environment. If the control software notes that relevant events are occurring ever more frequently, it can decrease the spy-hopping intervals. Sampling the environment more often will help guarantee smooth operation, but at the expense of increased energy consumption. When the robot's control software senses that external activity is slowing down, it can increase the spy-hopping intervals again to save power. This technique can be used in situations where the environment changes in a relatively predictable way. If the adaptive control software alters the spy-hopping interval fast enough to keep track of the changing environment, then all will be fine. If the environment changes faster than the control system can adapt, problems will arise. The spy-hopping interval may remain too large to effectively sample the sped-up environment. If limits exist on the rate of change in the environmental processes, then the adaptive control system can be designed to keep up. But if no definitive boundaries exist for these processes, be wary of adaptive control loops within the robot. It might be better just to waste the extra energy and let the control system run at the fast rate, rather than risk a control system problem.

Software Considerations for Energy Control

As discussed before, only hardware can conserve energy since it's the only consumer. Most of the hardware features capable of conserving energy will probably be under the direct control of the software. Many techniques for using software to save energy bear mentioning.

OPERATING SYSTEM

We've already discussed some of the *operating system* (OS) hooks that can be used to conserve power. By and large, the very presence of an OS is antithetical to the proper functioning of a parsimonious energy conservation system. We won't discuss this much further since each OS will have its own documentation for such matters, but be careful that the OS properly supports the energy conservation states of the processor that is running the OS. If the OS has not been properly ported to the processor, or if the OS does not support energy conservation, then consider another one.

One of the key features of an OS that we've mentioned is the handling of the software environment. The OS must be capable of storing and retrieving software environments so it can survive power failures. During the initial system engineering of the robot, we must decide what the implications of power failures are. If the robot must be able to survive a brief interruption of power, then special hardware and software considerations must be made. We'll discuss these in the section on power failures.

ALGORITHMS

We can tailor algorithms to conserve power. The central idea is that each individual operation in a control algorithm, each and every executed instruction step, consumes power. We can alter the algorithm so fewer steps are required and thus save power. Certainly, algorithms can be structured many different ways, but to save power, keep them short and sweet.

SCHEDULING

To coin a phrase, one should better buy the pizza instead of just eating it by the slice. Certainly, buying the whole pizza at once will be cheaper, and the same is true in the software domain. It becomes easy to chop a control problem into tiny pieces without realizing it. Often, this happens during the design process as various aspects of the power control problem are considered one at a time. Once a problem is chopped into pieces, we wind up paying for it in lost compute time, lost energy, and lower reliability. Problems become fragments in more than one dimension. Here are a few ways to decrease wasted overhead in the robot:

- **Computer overhead**   A computer control program that executes intermittently is wasteful. The robot's computer must be awakened or used more often, and the attendant overhead becomes excessive. If we can find a way to pull the program back together so it can be handled in one fell swoop, we can reclaim the lost energy and time. Consider auditing all the tasks the robot performs and identifying those that are being handled in a fragmented way. Several such tasks creep unnoticed into a design during the design phase. Rewriting those tasks will often bring power savings and make the software more reliable.
- **Power overhead**   Most robots have dozens of tasks to perform. Some of the energy to perform these tasks will be wasted in overhead. Consider, for the moment, a car. Starting a car, at the very least, causes energy to be expended from the battery. If we have two errands to run, we could group them together so we only have to start the car once. The same grouping technique can work in energy management in a robot. Some of the hardware circuitry will need to be charged up to perform tasks. We can save energy by grouping tasks together in time so less energy overhead is wasted.
- **Pipelining (real time)**   Consider modifying the robot's control software to pipeline tasks. To illustrate why this is useful, we need to revisit pipelining as it applies to processors.

In processors with pipelines, instructions are not executed immediately, but they are put into a pipeline. Pipelines can be used in different ways to control energy consumption or execution speed. A trade-off takes place between speed and power since more compute power requires more energy.

Pipelines for Speed

In Chapter 3 on computer hardware, we discussed using pipelines to speed up processing. The specific example used was "If A, then B, else C." In a pipeline optimized for speed, A, B, and C are put into the pipeline simultaneously and are computed simultaneously. Either B or C comes out of the pipeline (already precomputed) based on the value of A. This is the way a pipeline optimized for speed would behave. It burns energy as fast as possible.

Pipelines for Power

As the processor goes about executing the instructions inside the pipeline, it sometimes notices that some of the instructions don't have to be executed. Power can be saved if unneeded instructions are not executed. Let's revisit our example program, "If A, then B, else C." In a pipeline optimized for power, A is put in the pipeline and computed first. Based on the value of A, either B or C is put into the pipeline for computation. This method is clearly slower but saves the energy that might be used in unneeded computations.

We just looked at how a pipeline in a processor can be optimized for energy conservation. The processor has a pipeline that handles instructions that are executed in a serial manner. In the same manner, we can construct a pipeline of tasks that the robot executes in a serial manner. If we buffer up these tasks instead of executing them immediately, we may discover tasks that do not have to be executed. In a real operation, various commands may arise that just don't make sense. One set of commands might look like this: "Go to from Point B to Point C and pick up the Rock C. Bring it back to Point B and examine Fact A. If A is true, drop Rock C and pick up Rock B."

A properly constructed robot task pipeline would look at this series of commands and alter it to the following: "Examine Fact A. If A is true, pick up Rock B, or else go to Point C and pick up Rock C."

A very well constructed robot task pipeline would question whether the robot should do any of this work. If neither the information about fact A nor the rocks are needed in subsequent tasks, all this work can be avoided. If a subsequent robot task requires Rock B or Rock C, then the pipelined tasks can be executed. Further, the robot task pipeline can determine if the robot really needs to return to Point B at all.

Most people, while cleaning house, will find lots of reasons to go upstairs and downstairs to achieve specific goals. If no emergencies occur, it makes sense to pipeline all the tasks for a while. Go upstairs for the upstairs tasks and downstairs for the downstairs tasks. It's easy to tell that this saves energy. If the robot can afford to hesitate for a while, it can pipeline its tasks and probably save some energy.

Pipelining (Premission)

Just as the robot's computer can pipeline tasks, so too the robot designers can pipeline tasks well in advance. It's just a matter of how soon the logical order of execution can be determined. In the case of real-time pipelining, the robot's computer pipeline is stripping out tasks that have been cobbled together at the last moment. The real-time pipeline optimizes tasks that don't make sense because they could not be predicted beforehand. But with clever programming, the robot's designers can also optimize ahead of time the ways in which tasks are executed. It's almost like performing pipelining well before the tasks are to take place, and then feeding the robot's real-time pipeline a stream of tasks that don't need any further optimization.

Consider a trivial example. Suppose the robot has "shoes" that are required for movement. It does not take a genius to realize that putting on shoes should be done before standing up. Those of us with kids, however, know the kid's already up, in the car, and down to the mall before we discover the surplus of pink wiggly toes. Given that humans are leaps and bounds ahead of robots in their abilities and evolution, we leave it to the reader to discover the advantage this sort of behavior conveys to the human species. Why is the world put together like this? Once we, as humans, become smart enough to discover the reason, we will surely build superior robots.

But I digress. The robot designers should be able to plan missions where the robot is controlled well enough to put its shoes on before moving. In fact, with the proper development software, the premission planners should have the tools that will make proper robot control largely an automatic occurrence.

Taking one step back, robot designers should also be able to optimize all the software instructions to conserve power. We've already seen the example of the IF instruction optimized for speed or power. Most compilers are capable of optimizing the software for various things. With certain flags set at compile time, a compiler can turn out fast code or condensed code that uses little memory. A good compiler will also eliminate code that will never be executed.

SAFEGUARDS

The robot's control software should have control loops that will sense the inordinate consumption of power and other serious situations. This is especially vital in space missions or when the robot cannot be repaired. Two types of events should be watched carefully with separate software watchdogs:

- **Security breaches**   Communication coming into the robot should be scanned for evidence of hackers and other more random interference. If it's determined that the system is under attack, it should move to a safe configuration and shift its control strategies. The robot should report the intrusion once it's detected, and then secure the robot's energy supply against unwarranted use. Energy can be conserved while proper communications are restored.

■ **Power thrashing** Given that the energy supply is of critical importance in many mobile robots, it makes sense to observe the power drain carefully. If the energy is being drained away too quickly, it makes sense to shut down activities until the cause can be determined. The robot may be thrashing about, malfunctioning, or just executing a badly designed algorithm. It's a smart robot that will give itself a timeout.

POWER FAILURES

One technique that is all but lost in today's complex world of computer software is the use of power failure detection. It is possible to build a power supply with an output signal called *Power Failure Detect* (PFD) that will warn of the impending cessation of input power. During a power failure, the PFD signal can go low a few milliseconds in advance of the time when the regulated power will fail to meet specifications. The processor will be interrupted and can do all the housekeeping necessary to survive the event. If the robot is designed from the start to take advantage of this, then it is possible for the robot to pick up right where it left off. To plan on using this capability, we must solve all the following problems:

■ The power supply must generate the PFD signal reliably. Most power supplies do not have this feature.

■ The OS software must facilitate the implementation and use of the PFD signal. The truth is, most OS software will simply get in the way of successfully implementing PFD software. Most large OS software products have so many holes and gaps that success is problematic.

■ The robot's computer must have sufficient nonvolatile memory to put away all the volatile data that will be lost during the power failure. Flash memory, battery-backed *Random Access Memory* (RAM), and disks are all good places to put the data. Once a PFD is signaled, however, we must be very careful to finish all operations before the power fails completely.

■ All the robot's states must be put away to accomplish a complete PFD recovery. These states include both the digital states that we have been talking about and mechanical states. The robot, after all, may be moving when the power fails. It is likely that the movement will be disturbed by a power failure unless the power failure is very short. Consider the case where the robot is moving its arm to the right.

If a very brief power failure occurs, proper PFD software will preserve that information and finish the movement when the power returns a few milliseconds later. Certainly, if the power failure lasts longer, the motion will be ruined anyway. In addition, if safety requires it, all motions must come to a fail-safe stop when the power goes down. In all these instances, a complete PFD recovery of mechanical states is impossible.

Mechanical Considerations: Software for Energy Control

We'll be discussing some mechanical engineering in Chapter 11. Many aspects of the mechanical design of the robot hinge on the energy consumption. Certainly, if the robot moves, then energy is expended to create that motion. The control system can monitor the expenditure of mechanical energy and optimize things. This can happen in several ways, which are listed here in no particular order of importance.

SHARING MOTORS

Motors tend to be among the heaviest of components. If the robot does not have to move in multiple dimensions at once, consider putting in lightweight clutches and share the motor between these mechanisms. The robot's software may have to determine which direction to move first.

POSITION PREDICTION

When the control software decides to move the robot, it expects it to wind up in the proper position at the end of the move. But the truth is, the robot rarely winds up in the exact right spot after an initial move. Another smaller movement is often necessary. To the extent that these smaller corrective moves can be minimized, the robot can save energy. Remember, it often takes extra energy just to get a robot moving at all. If the robot's control system is smart enough to adapt, it can predict the effect of a movement even before it takes place. Further, as conditions change around the robot, the prediction mechanism can be altered to fit the conditions. With the right algorithm, the robot's control software will continue to be efficient in its movements, coming close to the predicted position on the first try.

Consider a real example. Suppose the robot must put fence poles in the ground. The control software has been turning on the forward motor for three seconds each time the robot must move to the next pole. However, as the robot begins to enter sandy soil, traction becomes a problem and it takes extra motor time to reach the next position. The control software should be able to sense this from the last fence post, and turn the motor on longer when moving to the next post. As the traction gets better, the duration can be decreased.

MINIMUM ENERGY ROUTES

The control system software, given a command to move the robot in multiple dimensions, should be able to minimize the amount of energy required to make the motions. This can take place in multiple ways. In some cases, the robot can effectively make the required motion in any number of different ways. Suppose, for example, that the robot must move its hand to a new location to perform a task. The robot could retract its hand, move itself to a comfortable spot in front of the object to be manipulated, and extend its hand to grasp the object. This set of motions might well be wasteful. Moving the hand to the required position may only take a rotation at the waist or an extension of the arm. The same task can be carried out in this manner at a great savings in energy.

The control software can decide which movement will minimize energy consumption in a few different ways. The software can contain a simple static model of the cost for moving in each dimension, or it can adaptively change the movement costs by observing the costs of previous movements. Certainly, these algorithms can become complex. If one portion of the robot breaks, rendering motion in one dimension impossible, simply raise the cost of motion in that dimension to a very high value. The energy minimization software should then bypass any movement in the dimension containing the broken components.

BRAKING

Anyone who has driven down a very long, steep hill knows that braking takes energy. The brake pedal is held down, requiring energy from the leg muscles. Common driving lore holds that the brakes should be let up now and then to avoid overheating. This is something I still do to this day, not knowing if it's needed. In any event, the design of the braking system should be carefully done instead of waiting until the last second.

First of all, just what are brakes? We'll discuss the types of brakes shortly. Defined in a general manner, brakes are a mechanism for slowing down the robot in one or more dimensions. Following the theory that every component must be justified, we should ask the following question. Why might braking be required at all?

Safety

If the robot gets in a difficult situation, it may have to stop quickly. This can occur if an obstacle appears, a malfunction occurs, or operators press the panic button. Note that in the case of a panic, brakes might actually hurt instead of helping. Consider the case where someone has become accidentally caught in moving mechanisms. Once motion is halted because of a panic, the brakes should be released as long as no more motion ensues. With the brakes released, the mechanisms may be moved to extricate a trapped operator. In designing the robot, don't forget that the brakes can be as dangerous as the motors.

   The control system software to deal with braking is a lot more sophisticated than it might seem at first glance. Consider for the moment *antilock braking systems* (ABS) in cars. When the computer that runs ABS senses a skid, it pumps the brakes to help keep the car skidding in a straight line and to maximize brake's gripping action. Here's an article on ABS using fuzzy logic, if a fuzzy braking system appeals to you: www.intel.com/design/mcs96/designex/2351.htm. Some more good articles on ABS can be found at www.howstuffworks.com/anti-lock-brake.htm and www-s.ti.com/sc/psheets/slit114a/slit114a.pdf. Some engineers spend their entire careers in this field.

Power Failure

If power fails, the robot may go out of control. What happens next depends on the brake design. Cars have two kinds: temporary brakes (the operator can press the brake pedal) or flip-flop brakes (the operator can pull the emergency brake and release it later). A third option would be automatic braking on power failure, where the brakes are kept off until the power fails. The astute robot designer must choose between these options. Control system software will only be of use until the power completely fails. If the robot's power supply has PFD built in, some warning will be given in advance. Although the primary braking system can become complex, keep the emergency braking systems dirt simple.

Speed

The fastest way to go from point A to point B is to accelerate at the maximum rate for half the journey, and then decelerate at the maximum rate for the other half of the journey. Those well versed in calculus will recognize the several flaws in this last statement, but it gives us the basic concept. If speed of operation is the goal (instead of energy conservation), then techniques such as this braking maneuver can be used to decrease travel time. We leave it up to the reader to work out the math model involved to truly minimize the overall trip time.

What Types of Brakes Exist?

Remember the general definition. Brakes are a method of slowing down (or remaining in place). This is a function that can be implemented in the following ways:

- **No brakes** Okay, we've all had bicycles like this. The truth is, aside from scraping shoes on the ground, it's possible to slow down just by coasting to a stop. This does not work real well going downhill, but it works just fine on level ground and going uphill. Even if the robot has great disk brakes, the control software should be smart enough to recognize when they don't need to be used. This sort of braking action consumes very little energy, but it requires rather sophisticated software. Here's an example of the type of software action that could save energy. Suppose the robot must move 4 feet. Suppose from experience the robot knows it will coast 2 feet once the robot is at top speed and the motor is turned off. It's likely that the least energy-expending method of moving is to get to top speed, move for 2 feet, turn off the motor, and coast for 2 feet until the robot comes to a stop. Other power expenditure plans may work better, but certainly little power will be wasted in the last half of the journey. The motor and the brakes will both be off. One thing is for sure though. The robot will not complete the move in the minimum amount of time.

- **Motor braking** Just as a motor can be used to accelerate a robot, so too can it be used to decelerate. Motors can be used as brakes in a couple of different ways. Because moving coils of wire through magnetic fields cause a current to flow, some motors become generators when the rotor is spun around. If the motor coils are shorted out, then a larger current will flow and the motor will resist the spinning motion on the rotor. By definition, this causes braking. More sophisticated motor control circuits are available that can brake more effectively by driving the motor coils in an optimum fashion. In fact, the motor can be partially driven in the opposite direction. The motor then actively counters the robot's existing motion.

- **Pad brakes** Regular friction brakes of all sorts are available too. We've already discussed ABS brakes and the various forms of braking actions (manual and automatic). It just makes sense to mention them again here. However, one thing hasn't been mentioned before. Brakes require cooling. In the worst case, they dissipate the entire kinetic energy of the robot. Providing for the cooling of the brake pads (if they exist) must be part of the design.

TORQUE CONTROL

Much like ABS brakes can prevent spinning wheels from locking up, it makes sense to prevent wheels from spinning during acceleration when they should be gripping the traction surface. It does no good to spin the robot's wheels when it is accelerating. That's just a waste of power, time, and rubber. (The tire makers in Detroit will be glad I cannot conceive of moving on anything other than tires.) The following discussion assumes the robot has more than one speed or can choose between more than one torque setting on the wheels. To counteract spinning wheels, the robot must first be able to sense the event. The robot's control system can sense when the tires are spinning in several ways.

The simplest method is to determine the speed of the robot over the terrain and compare it to a model of the wheels. If one wheel is spinning significantly faster than the others, it is probably not gripping the same surface. The same sensors used in ABS brakes would work in this case.

A slightly more difficult method is to sense the torque on each wheel directly. This can be done with spring mechanisms or by monitoring the voltages on the motor windings. A motor meeting no resistance will not consume as much power to spin the wheels at a known rate. If the wheel is spinning, the motor control circuitry should be able to signal that.

RECLAIMING ENERGY

One of the features that comes almost for free with an electric car is the ability to generate electricity when going downhill or braking. (A fun web site that should come in handy and that details much of the thinking that has gone into electric cars is at www.howstuffworks.com/electric-car.htm.) If a robot takes 100 watt-hours of energy to climb a hill, we might think we could reclaim most of those 100-watt hours by going down the other side of the hill. But alas the laws of thermodynamics get in the way. Surely, we would not want the thermodynamic police to be on our tail.

The second law states that the entropy of an isolated system can never decrease. This limits the efficiency of energy conversion between different types of energy. It's rarely possible to approach 25 percent efficiency converting electrical energy to kinetic energy and back to electrical energy again. Reclaiming energy is very difficult and should only be attempted if the equipment is virtually free and does not interfere with other processes. It rarely pays off in a device as complex as a robot. More info on thermodynamics and energy conversion can be found at http://members.aol.com/engware/systems.htm.

Revisiting technology is one of the pleasures of writing a book like this. During my search for good supplementary web sites, I often run across some odd twists on things. For some truly interesting reading, I offer the satirical web page of the Thermodynamic Law Party (http://zapatopi.net/tlp.html). The thermocrats among you will already recognize the principals therein. For the rest of us, read this site with care. On the site, it states that Kelvinian meditation causes epileptic seizures "only in lab mice at extreme doses." At the very least, that should prod the curious. As in all things, some truth can be found in everyone's thinking.

ENERGY REUSE, REVISITED

Although it is difficult to reuse energy by converting it from one form to another, it is easy to reuse energy in its existing form. We've already seen how we can use the existing kinetic energy of the robot to coast to a destination and save energy. We can extend this concept further by keeping track of the kinetic energy in various parts of the robot. Here's an example.

Suppose a robot has a relatively human form. This being the case, we can run a quick experiment using on our own bodies. Stand up one arm's length away from a light switch on the wall with your left shoulder closest to the wall. Now turn so that your right shoulder is closest to the switch with your left shoulder away from it. If you want to turn on the light switch with your left hand, you have a couple of ways to accomplish this task.

You can rotate right (90 degrees) at the waist until facing the wall and only then raise your left arm to touch the switch. These two motions are disjointed and consume relatively known quantities of energy.

An alternative way to do this is to raise your arm to touch the switch when the rotation is halfway completed (45 degrees). It may seem easier to do it this way because the momentum of the arm is already headed in the direction of the switch when the rotation is halfway completed. But if the rotation of the waist is completed before the arm is raised, energy is wasted in raising the arm.

The bottom line is that robots can use coordination. Very few people ever bother to define just what human coordination is. All we know is that some athletes seems to soar above the others effortlessly and perform dazzling feats. But broken down to physics, at least some aspects of coordination come down to energy conservation and the conservation of momentum. Just as the human brain must act to turn a awkward person into a graceful athlete, so too a robot's control system must run algorithms capable of streamlining the motions of the robot.

The motion and energy computations that would streamline the motions of the robot need not be done at the spur of the moment just before they are needed. It is possible to compute many of the motions ahead of time and store the results for future use. The designers of the robot can experiment in advance to find the proper combinations of motions to achieve a desired effect. If the robot's repertoire of motions is small, this may work well. But if the robot must move in multiple dimensions at once to achieve complex, spur of the moment tasks, then the control system may need to perform these calculations quickly, in real time.

Writing a software program to simulate coordination is a complex task. A good, first-order approximation would be to write separate control algorithms for each component. For example, we can write one control loop for the arm and one control loop for the waist. While the control loop for the waist is rotating toward the wall, the control loop for the arm will recognize the optimum time to start moving the arm.

It is possible to run into some trouble with many control algorithms running in parallel, but these difficulties can be overcome. Detecting and avoiding hazards, for instance, can become a problem. Moving one component at a time is more predictable because only one control loop is active at a time. If the waist and arm control loops are both operating at the same time, they must be coordinated if obstacles must be avoided. Coordination involves communication and falls prey to all the difficulties we discussed previously in parallel processing. If we watch the pitfalls, we can reap the rewards in energy savings.

Another example of coordination involves the rotation of mass. Ice skaters pull in their arms when they go into fast spins. A robot that must rotate should pull in its arms before the rotation. Not only does it help avoid punching the operator, but also less rotational energy is needed.

A good article on designing a low-power system is at www.iapplianceweb.com/story/OEG20020623S0006, and a review of some of the electrical engineering techniques we've discussed can be found at http://academic.csuohio.edu/yuc/talks/low-energy2k1021.pdf.

Another interesting article can be downloaded from wwwhome.cs.utwente.nl/~havinga/thesis/ch2.pdf. The author clearly views the world in terms of energy. Table 3 in this article seems to indicate the average human expends daily the energy equivalent of a kilogram of coal, or roughly the energy in 10 beers. Check the chart out; it might explain some of the neighbors!

Bottom line, the conservation and control of the robot's energy reserves requires great care. Software algorithms, property written, can minimize the robot's consumption of energy.

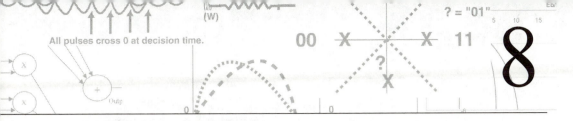

8

DIGITAL SIGNAL PROCESSING (DSP)

All humans practice *digital signal processing* (DSP) daily. This may come as a surprise, but it's true. Further, very few people know the simple theory that they actually practice each day by instinct alone. In this chapter, we'll discuss the theory and relate it to real-life examples.

First, let's quickly review how DSP functions. Most of the real world is analog, not digital. The robot will need to look at signals of all sorts. These signals have to be accessible to the control computer so the proper processing can occur.

Figure 8-1 shows one way this can be done. An *analog-to-digital* (A/D) converter digitizes the analog input signals. The digital representations of the signals then go into the computer where they are processed as needed for the application. The computer can then output digital results, some of which can drive a *digital-to-analog* (D/A) converter, which generates analog signals for output. Each element in this chain of electronics serves to modify the information from the original signals in various ways. We'll discuss the characteristics of each block in the figure later in the chapter, but for now, just realize that the computer cannot see the analog signals at all times. It can only sample

191

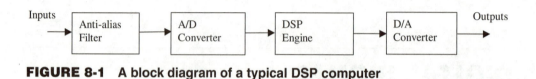

FIGURE 8-1 A block diagram of a typical DSP computer

them periodically with the A/D, and it has no idea what the signals do between samples. We'll state the main theorem used in DSP and then demonstrate that we already know the theorem and use it instinctively every day.

The Nyquist-Shannon Sampling Theorem

We cannot capture the essence of a digitized signal without sampling it at a frequency twice that of the signal. Stated another way, we must sample a signal twice as fast as the highest-frequency component in the signal.

ANTI-ALIASING FILTER

To successfully sample a signal, we must first alter it to filter out all the frequency components that are above half the sampling frequency. The frequency at 50 percent of the sampling frequency is also called the Nyquist Frequency. We'll get into a discussion about just what aliasing means later. These statements are oversimplifications of the original theorem. Consult the URLs near the end of this section for a more thorough treatment.

So where do we use all this math theory in our daily lives? Here's one for readers with kids. Nobody pays constant attention to the kids. It's impossible to do so because it takes too much energy and, further, paying constant attention teaches them nothing. Instead, we sample their behavior periodically by listening in on them. Often we turn our heads, cup our ears to listen, and say, "Gee, it's way too quiet up there." Oddly enough, with kids, the total lack of input is the very signal that something is wrong.

That was an easy example. Here's a harder one. Consider the following experiment —don't do it for real. While you are a passenger, just imagine you are driving and paying attention to the road. Drive down the street past a long row of parked cars. At a constant speed, pass one parked car each second. It's not possible to watch every car every second. The truth is, we sample the road ahead with our eyes.

So here's a question. How often must we sample the parked cars to feel comfortable about driving by them at this speed? Remember, we are driving past one car per second. Let's assume we close our eyes and only open them briefly at a fixed sampling rate. How often do we have to open them to feel comfortable?

Well, to confess, I tried this stupid experiment. It's a little bit like a doctor injecting himself with germs to test out his new vaccine. I did it safely though. Here's my report. Keeping my eyes closed was intensely uncomfortable, and I didn't try it very long, which was certainly to be expected. Opening my eyes once a second was uncomfortable. I could only see each car once as I passed it. Opening my eyes twice a second was more comfortable in that I felt I could control the car properly.

In this experiment, I experienced the Sampling Theorem firsthand in a conscious manner. To observe the cars properly, I had to sample the cars twice a second in a situation where the cars were going by once per second.

Critics of this experiment might say, "That's great, but what if a fast-moving car came darting out of a side street? Wouldn't that cause an accident?" The answer is yes. Sampling might not work properly if an unexpected car appeared on the street. If we got lucky, we would notice the fast car when our eyes were open and we might be able to avoid it. We would probably not be able to tell how fast it was going though. Worst case, we would never even see the fast car; it would both appear and hit us while our eyes remained closed.

The key here is an antialias filter, which, in our example, would be a speed limit sign. Town planners automatically protect the quiet side streets (those with rows of parked cars) by surrounding the neighborhood with speed limit signs. The fast-moving vehicles are therefore filtered out of the situation. If fast-moving cars were the norm in the neighborhood, we would be on guard and sample the road ahead much more frequently. We react instinctively as we apply the Sampling Theorem in this way.

Let's summarize the driving experiment in DSP terms. Cars are driven at all different speeds; these are our input signals. To protect our sampling system, we put in an antialiasing filter (speed limit signs) so we do not have to deal with cars moving faster than one car length a second. Driving past parked cars at one car per second, we sample the cars visually two times a second. Per the Sampling Theorem, this gives us enough information to process the data and to drive carefully.

Let's try another experiment. We will use pure sine waves as input signals to the DSP system and will sample at a fixed rate every 0.3 seconds. This works out to a sampling rate of 3.33 Hz or roughly 20 radians per second. We will vary the frequency of the analog input signals from 3 to 15 radians per second. With a fixed sampling rate of 20 radians per second, the Sampling Theorem predicts we will do a good job of sampling sine wave input signals with frequencies as high as 10 radians per second. By looking at sine waves from 3 to 15 radians per second, we should see a breakdown in the sampling systems above 10 radians per second. We have, after all, eliminated the antialias filter from the DSP system to illustrate the problems that could occur in its absence. We should expect problems.

Take a look at the evidence in the following figures. Each chart pair shows the input sine wave on top and the sampled result on the bottom. These charts were made in a spreadsheet, which attempted to fit a curve to the sampled data at the bottom. The waveform thus reconstructed from the sample data is shown on the bottom of each chart. It represents what the DSP computer thinks the original waveform looked like (see Figure 8-2).

The sampling went reasonably well from 3 to 9 radians. Looking at Figure 8-2, it's clear the software could not discern the frequency (or the shape) of the input sine waves with frequencies above 10 radians per second, but something else emerges. The sampled waveform looks increasingly like a lower-frequency signal. Take a look at what happens in Figure 8-3 as we extend the charts well beyond a 15 radian per second input signal. The sampled waveforms seem to decrease in frequency from 16 through 21 radians per second, and then increase in frequency again between 21 and 26 radians per second. The sampling system thinks the real waveform is doing something that is is not doing. This is classical aliasing right before our eyes. The sampling system is being fooled.

An alias, as defined in *Webster's* dictionary, is an "assumed name." The sampled, reconstructed waveform at 16 radians per second looks like a waveform only two-sevenths the same frequency. It's representing itself as something it is not, hence the name alias.

We've all seen this exact same effect take place with car wheels. At night, under incandescent lights, look at the hubcaps of a moving car as it slows down to a stop. Pick a car with many spokes in the hubcap. Because electrical power is at 60 Hz (or 50 Hz elsewhere), electric lights flash at that frequency. The lights are effectively sampling the hubcap spokes for our eyes. We can only see the hubcaps when the lights are at their brightest. As the car decelerates from high speeds, the hubcaps appear to slow down to zero before the car has even stopped. Then, as the car continues to decelerate, the hubcaps appear to start moving backwards. This is the exact same effect that we just saw in our charts about aliasing.

To avoid having the DSP computer fooled in the same manner, pay strict attention to the Sampling Theorem. Have the computer sample at twice the highest frequency in the input signals. Further, put an antialiasing filter in the input of the D/A that will filter out all frequencies above half the sampling frequency.

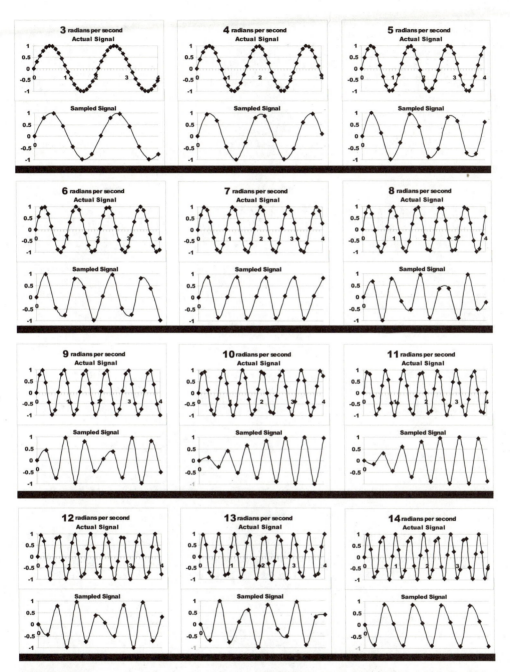

FIGURE 8-2 Sampling Theorem example: When sampling at 20 radians per second, things break down for signals faster than 10.

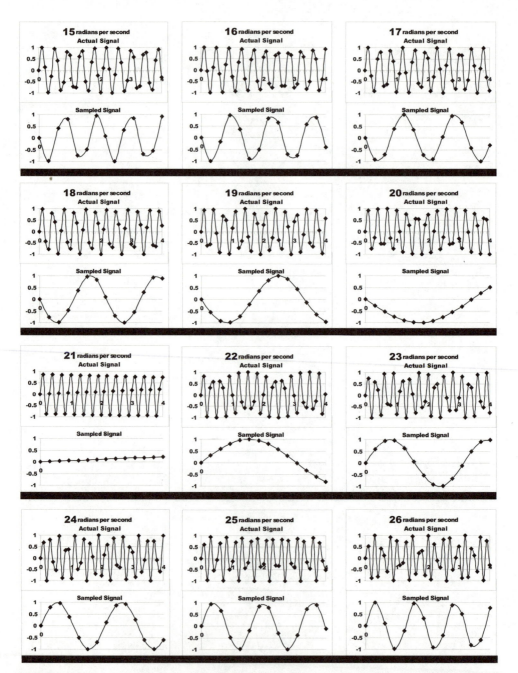

FIGURE 8-3   Aliasing example: When sampling at 20 radians per second, aliasing is evident past 10 and dramatic by 20.

Here are some further descriptions of the Sampling Theorem:

- http://ccrma-www.stanford.edu/~jos/r320/Shannon_s_Sampling_Theorem.html
- http://ptolemy.eecs.berkeley.edu/eecs20/week13/nyquistShannon.html
- www.hsdal.ufl.edu/Projects/IntroDSP/Notes/Sampling%20Theorem%20Brief.doc

If you want to have some fun with language, take a look at the www.nightgarden.com/shannon.htm web site.

With such great theorists like Nyquist and Shannon being brought up, I feel odd about injecting some practical details into this discussion (see Figure 8-4). Unfortunately, it has to be done. The world is a tough place, Grasshopper, and one cannot go about spouting generalities without getting in trouble. So hold your nose; here comes some castor oil!

DSP is all about transforming data so it can be processed and used to good effect. The trouble is, most of the transformations distort the data along the way. Before we even get started with DSP, we find that the antialias filters and the A/D both alter the data in ways that must be carefully taken into account. Further, once the DSP processor and the D/A come into play, we will see that they too distort the data.

It's all very easy to slap an A/D and a D/A onto a computer and call it a DSP system. The difficulty comes in making it see the world correctly and helping it make the right decisions. So here are some of the salient details that should be taken into account.

FIGURE 8-4 Nyquist and Shannon

A/D Conversion

We're not going to discuss the types of A/D converters that are available, nor are we going to discuss how they work. We leave it up to the reader to delve into these details, including cost and linearity. Just remember that it must be fast enough to keep up with the sample rate chosen according to the Sampling Theorem. Here are a few good URLs that talk about A/D conversion in general:

- http://hyperphysics.phy-astr.gsu.edu/hbase/electronic/adc.html
- http://jever.phys.ualberta.ca/~gingrich/phys395/notes/node151.html
- www.sxlist.com/techref/io/atod.htm

We do need to have a discussion about the number of bits in the A/D. First of all, we must recognize that an A/D converter's primary characteristic tends to be the number of bits in the digital output. Be wary of A/Ds that have many bits. It's not unusual for an A/D to fail to perform up to its reported level. So even if an A/D touts 16 bits of resolution, it may only deliver the equivalent performance of 12 or 14 bits. It seems obvious that a real-world signal cannot be well represented by just 2 or 3 bits of data. But how many bits do we really need?

First, we need to define db or decibel. This acronym has many uses, which each have their own definition. Here we will take it to mean a method of measuring voltage ratios. A voltage signal that is 6 db lower than another is just 50 percent of the other. Increasing a voltage signal by 6 db doubles it. In a similar manner, 20 db connotes a factor of 10. A good web site on decibels is at www.its.bldrdoc.gov/fs-1037/dir-010/_1468.htm.

The primary consideration when looking at A/D bit length is the nature of the input signals. What *signal-to-noise* (S/N) ratio do the signals have? All signals have noise on top of them. Some signals have far more than others. If a signal is roughly 10 times bigger than the noise, then it is 20 db S/N. Figure 8-5 shows a visual representation of noise at different S/N ratios.

It's important to know the S/N ratio of the signals being measured. The rule of thumb is that each extra bit in the A/D provides another 5 db of S/N capability in the DSP engine. Ordinarily, another bit would double the effective range of a word and thus provide 6 db of S/N capability, but I've been told by experts not to expect the theoretical limit, so count on 5 db per bit.

Now if the signal to be measured has a 40 db S/N ratio, then an 8-bit A/D might be just the ticket since $8 \times 5 = 40$. As long as stepping up to a couple of more bits is not too expensive, I'd consider a 10-bit A/D for such a job. Buying a 16-bit A/D will *not* convey any extra accuracy with such a low S/N signal. Ordinarily, a 16-bit A/D might allow 80 db of S/N processing (5×16), but if the input signals are not up to that num-

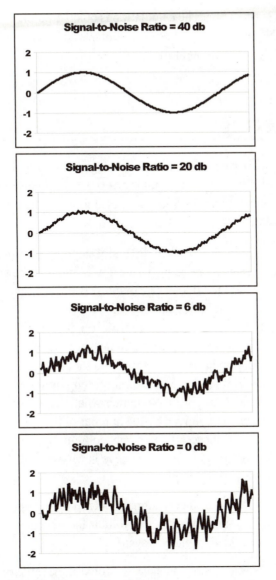

FIGURE 8-5 A visual look at S/N ratios

ber, there's no sense trying for more. In general, use an A/D that's just somewhat better than the signals it must measure.

So here's our first pop quiz! If the signals have an S/N ratio of 60 db, how many bits of resolution should the A/D have?

It should have at least 12 bits. The calculation is 60 db/5 db/bit = 12 bits. More information on the S/N ratio can be found at http://searchnetworking.techtarget.com/sDefinition/0,,sid7_gci213018,00.html.

A/D Dithering

A/D converters are not perfect. They convert analog signals into digital representations of the original signal. If the original signal is a very smoothly changing signal, then the digitization of the signal can add significant noise to the signal. This comes into play in at least two situations:

- Sometimes the A/D itself will have difficulty stepping over major bit boundaries. Suppose, for example, we're using a 16-bit A/D and that the signal steps over the boundary from 7FFFH to 8000H. The number 7FFFH is in hexadecimal (base 16) notation explained at the following URLs:
    - www.whatis.techtarget.com/definition/0,,sid9_gci212247,00.html
    - www.hostingworks.com/support/dict.phtml?foldoc=hexadecimal.
  Many bits are changing at the same time, and the A/D may have trouble keeping the same accuracy it might have with simply stepping from 7FFEH to 7FFFH.
- Quantization error also creeps in. No matter what, the A/D can only represent the input signal to the accuracy given by the number of bits in the A/D. In a smoothly changing input signal, these effects can become noticeable. This effect is most often seen in graphic images; the human eye is very efficient at picking out error patterns in smoothly changing pictures.

To counteract these effects, a random signal is added to the input signal. This dithering of the input signal is generally sufficient to blur the deleterious effects mentioned earlier. Dithering can be added in many ways:

- **Analog noise**   We can simply put a noise source at the input of the A/D. The magnitude of the noise source should be just about the size of the quantization noise. If the range of the A/D is 10 volts, and it's a 10-bit A/D, then a single bit change in the A/D digital output covers $10V/2^{10} = 10$ mv. Adding a 10 mv noise source to the analog input stage would create the type of dithering needed. Using a noise source larger than 10 mv would also work, at the expense of lower resolution.
- **Random shifting**   One way to get around A/D imperfections is to dynamically (and randomly) shift the range of the A/D. A random voltage can be added to the input of the A/D and later be subtracted digitally from the A/D output. All the con-

version hardware is thus operated at random levels within the operating range. A web site describing this method is www.chipcenter.com/TestandMeasurement/ tn024.html.

- **Digital noise**   This can be added to the A/D output. This technique is perhaps the easiest to perform and it can be done with hardware or within the DSP processor.

Here are some dithering web sites:

- www.cinenet.net/~spitzak/conversion/dithering.html
- www.audioease.com/Pages/Barbabatch/TechInfo.html#aDithering
- www.edi.lv/dasp-web/sec-6.htm

Sample and Hold (S/H)

Analog inputs might be changing when they are sampled. Even after filtering out the high-frequency components in the antialias filter, the input to an A/D might be changing while the A/D is performing its function. Some A/D converters might be thrown off by a changing input, delivering an erroneous output. If the A/D converter must have a stable input during the conversion process, then the converter itself generally has a *sample and hold* (S/H) amplifier built right into the A/D converter. If it does not, we would have to add one before the A/D input. The S/H amplifier has a hold input that controls the hold function. When low, the S/H amplifier's output simply follows the input. When high, it takes a quick snapshot of the S/H analog input value and freezes the S/H amplifier output at that value. The S/H maintains this value long enough for the A/D to convert it to a digital value.

Further information on S/H amplifiers can be downloaded from www.national.com/ an/AN/AN-775.pdf and www.om.tu-harburg.de/Download/Datasheets/Linear/NE_ SE5537.pdf. Check the application sections and the tips on acquisition.

Antialias Filters

Now that we've got some idea what has to be inside the A/D block in our DSP system, what about the antialias filter? Well, the news here is even a bit tougher. We made a statement a while back that the antialias filter should be a low-pass filter that filters out *all* frequencies above the Nyquist Frequency. The ideal antialias filter would pass all frequencies (untouched) up to the Nyquist Frequency. Above that breakpoint, the

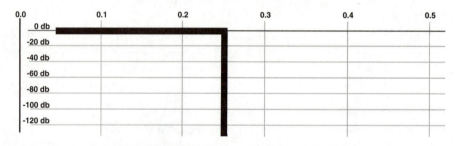

FIGURE 8-6   A perfect, but impossible to find, antialias filter

antialias filter should pass nothing. Figure 8-6 shows the nature of such a perfect antialias filter.

The figure shows the filter's response versus frequency. We can see that the filter perfectly passes all signals lower in frequency than 0.5 × Fs, the sampling frequency, which is 0.5 in this example. Above that point, the filter passes nothing at all. This chart is a typical frequency response chart for a component. The trouble is, it's impossible to build a filter that can do this. We must make compromises to achieve a suitable antialias filter design.

So what problems exist with designing the perfect filter, as shown in the figure?

EXPENSE

An ideal antialias filter with an infinitely steep rolloff (defined shortly) like that in the figure cannot be made. Filters are made with real-world components that have definitive, complex impedances. This means the filter will have a transfer function that reduces to differential equations with continuous solutions. This is all a complex way to say that the filter's frequency transfer chart will not have vertical rolloff lines. The filter must have curves and ramps. The vertical dropoff shown in the ideal filter will actually have to roll off with a less vertical drop. The more vertical the drop, the more expensive and complicated the filter must be. This puts us in a bind. If we want a more perfect filter, our expense goes up. If we want to save money, we will have to settle for a less perfect filter.

The typical solution is to put the antialias filter at a frequency a bit lower than the Nyquist Frequency and roll it off at a more gentle (cheaper) angle. A very similar solution is to put an imperfect antialias filter at the Nyquist Frequency and then move the sampling frequency up about 20 percent. We'll look at filter design shortly.

DISTORTION

The antialias filter itself will distort the very signals we are trying to measure. This occurs because most signals are a mixture of different frequency waveforms. Only pure sine waves contain single-frequency waveforms. Even a pure sine wave signal will get distorted some by a filter, but signals composed of several frequency waveforms will get distorted all the more because the different frequencies are treated differently by the filter. We will see that even distortion can be used to our advantage if the distortion can be predicted.

Over the years, the design of antialias filters has settled on a couple of good solutions that designers can live with. A good filter will have a steep rolloff and a deep stopband, as shown in Figure 8-7.

ROLLOFF

The rolloff is the slope of the frequency response between the passband and the stopband. With an operational amplifier and a couple of components like an inductor and a capacitor, it's possible to get a 12 db/octave rolloff. This means that for every doubling of the frequency, the filter attenuates the signals by a factor of 4.

STOPBAND

For a low-pass antialias filter, the stopband covers those higher frequencies that the low-pass filter is supposed to eliminate. The stopband is the area to the right of the rolloff curve that is dramatically lower than the low-pass frequency part of the curve.

As a rule of thumb, if the S/N ratio for the signals of interest is 40 db, we would want all the actual high-frequency noise in the stopband to be 40 db or better down in the stopband, such as Figure 8-7.

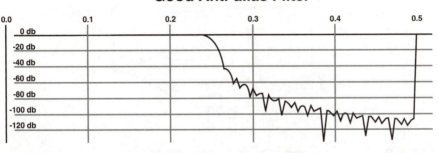

Good Anti-alias Filter

FIGURE 8-7 **An imperfect but realizable antialias filter**

ANALOG FILTERS

One simple way to make an antialias filter is with traditional analog electronics. With very few analog components, it's possible to get a filter with a decent rolloff and stopband. Figure 8-8 shows a schematic of a simple second-order filter and the transfer function that goes with it.

L is the inductance, C is the capacitance, and R is the resistance. Resorting to Laplace notations for the moment, the differential equation for this circuit is derived as follows:

$$Vout = ((1/Cs)/(1/Cs + R + sL))Vin$$
$$Vout = Vin/(s^2LC + RCs + 1)$$

This same calculation is carried out at the following web sites:

- www.ee.polyu.edu.hk/staff/eencheun/EE251_material/Lecture1-2/lecture1-25.htm
- http://engnet.anu.edu.au/courses/engn2211/notes/transnode19.html
- www.engr.sjsu.edu/filt175_s01/Proj_sp2ka/act_fil_cosper_fold/act_fil_cosper.htm
- www.t-linespeakers.org/tech/filters/Sallen-Key.html

The transfer function is shown in Figure 8-9. The rolloff of this circuit is 12 db per octave. Since this particular circuit rolls off indefinitely, the stopband should be well below the noise floor of the input signal (and thus not a factor).

We should recognize that the differential equation of this circuit is very similar to the second-order control system we studied in Chapter 2 on control systems. That's because direct analogies exist between the types of components as follows:

- Capacitors are the analog of mass. Just like energy is stored in mass as it gains speed, so, too, energy is stored in a capacitor as electrons flow into it and the voltage builds up.

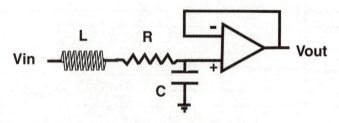

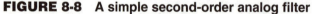

FIGURE 8-8 A simple second-order analog filter

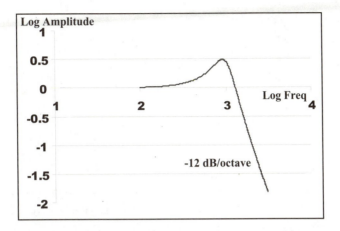

FIGURE 8-9 The frequency response of the second-order analog filter

- Inductors are the analog of springs. Inductors, like springs, act as an energy storage element. Current moves through an inductor, creates a field around the inductor, and builds up the voltage across it. Just like a spring can run out of stretch, so too an inductor can exhaust the magnetic materials that absorb energy to create the field around the inductor. As long as the amount of energy stored in the inductor stays below a certain amount, it will function properly. The same is true of a spring.
- Resistors are the analog of friction. A resistor, like friction, acts to slow down and drain off the movement of energy between the other two components in the circuit.

The filter's response to a step input is shown in Figure 8-10. The curve should look very familiar since it's virtually identical to the second-order control system we discussed before. The circuit could be used to drive a servo amplifier, but we leave it up to the readers to figure out, given R, L, and C, how to find the values of the damping constant d and the frequency v. It's not our business here to use this circuit for anything other than an antialias filter.

Given our example of a system with a 40 db S/N ratio, and using this particular circuit as an antialias filter, we can see what compromises we might have in the design of our sampling system:

- If we have a second-order analog filter with a 12 db per octave rolloff, we'd need better than 3 octaves to attain the desired rolloff for antialiasing:

$$(3 \; octaves \times 12 \; db/octave + 4 \; db) = 40 \; db$$

Amplitude

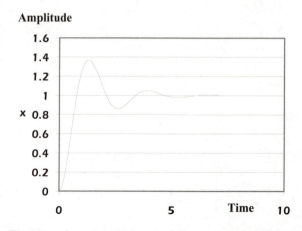

FIGURE 8-10   **The step input response of the second-order analog filter**

To get the stopband down to 40 db at the Nyquist Frequency with this filter, we'd have to increase the sampling rate by a factor of 10 or so (3 octaves +).

■ If we concatenate 2 such analog filters, we would get a 24 db per octave rolloff and it would only be something less than 2 octaves to achieve the same results:

$$(2\ octaves\ \times\ 24\ db/octave\ -\ 8\ db)\ =\ 40\ db$$

To get the stopband down 40 db at the Nyquist Frequency with this filter, we'd have to increase the sampling rate by a factor of 3.7 or so: (2 octaves −).
This would be a good trade-off since the analog filters are relatively inexpensive, and the DSP filters can be expensive, depending on the technology used.

■ If we concatenate 3 such analog filters, we would get a 36 db per octave rolloff and it would only be something more than 1 octave to achieve the same results:

$$(1\ octave\ \times\ 36\ db/octave\ +\ 4\ db)\ =\ 40\ db$$

To get the stopband down 40 db at the Nyquist Frequency with this filter, we'd have to increase the sampling rate by a factor of 2.1 or so: (1 octave +). This, too, would be a good trade-off. Details about analog filters can be found at http://my.integritynet.com.au/purdic/lcfilters.htm and at www.freqdev.com/guide/FDIGuide.pdf.

DSP FILTERS

There's no reason not to make an antialias filter using DSP techniques. We'll be discussing how to synthesize a DSP filter next. Here are some good web sites and a PDF file covering antialiasing filters:

- www.alligatortech.com/why_low_pass_filtering_is_always_necessary.htm
- www.dactron.com/pdf/appnote/aliasprotection.pdf
- http://kabuki.eecs.berkeley.edu/~danelle/arpa_0697/arpa.html
- http://members.ozemail.com.au/~timhoop/intro.htm

D/A Effects: Sinc Compensation

At the output of the DSP system, the D/A generates an output stream of analog values. The D/A only outputs a series of analog values that look like a rectangular staircase of constant voltages. Thus, the D/A inherently alters the output signal with the sinc function, which we'll discuss again shortly. What's needed within the DSP filter is an antisinc compensation filter.

This antisinc precompensation filter can reside inside the DSP compute engine. Let's say the DSP compute engine generates D/A output values at a rate of N per second. The antisinc predistortion computations are now added at the tail end of the DSP compute engine. Just how this is done is up to the designer. Since all these systems are assumed to be Linear Time Invariant systems, the antisinc filter can simply be added right into the middle of the DSP calculations. The previous D/A results are fed into this new compute block that runs computations for the antisinc compensation. The result is a new compute block outputting a stream of D/A values at a rate faster than rate N. The D/A will then run at a higher rate than normal. We smooth out the D/A values with a simple low-pass filter at the D/A clock rate. The resulting output waveform will not be overly distorted by the sinc effect. Note that running the D/A at a faster rate will mean higher energy consumption.

Here are some PDFs further discussing sinc precompensation:

- http://pdfserv.maxim-ic.com/arpdf/AppNotes/A0509.pdf
- www.lavryengineering.com/pdfs/sample.pdf
- www.ee.oulu.fi/~timor/EC_course/chp_1.pdf

DSP Filter Design

DSP filters are engines that do just exactly that: They process digital signals. DSP filters process digital data in an organized way. DSP can be accomplished in hardware *Field-Programmable Gate Array* (FPGAs) or the processing can be done in software. Even a general-purpose computer can perform DSP calculations. DSP filters are a mathematical construct that can be realized in various physical ways. We will discuss the mathematical structure first and the physical implementation much later in a separate section. Until we get to that section, none of the following discussion refers to specific physical implementations. This is a discussion in mathematical terms.

DSP filters process a digital stream that represents a signal. The stream of data will be recomputed in a coordinated way to form the output stream of the filter. It is the nature of the computation that gives the DSP filter the desired frequency transfer function. DSP filters can be constructed in many ways, but a few standard ways exist for building such a filter. A standard DSP filter is defined by its structure: a generic sequence of arithmetic operations executed on the input data stream. To make a custom filter, designers take a standard DSP filter and modify it. Tools and formulae convert the custom filter transfer function to a set of alterations of the standard DSP filter. The alterations, when made, turn the standard DSP filter into the custom filter. To actually construct the custom filter, the designers map both the standard DSP filter and the custom alterations to a physical implementation.

One of the simpler standard structures for a DSP filter is the *Finite Impulse Response* (FIR) filter shown in Figure 8-11. The data sequences through a linear series of registers called taps. At each sampling clock, the data moves to the next tap. After the last tap, the data is discarded. The output of the FIR filter at each clock is generally a single data element formed by combining all the data in all the taps. The data in each tap is multiplied by that tap's coefficient and the results are summed to make the output data. It is the vector of coefficients that turns the standard DSP FIR structure into the custom FIR filter. Once the designers decide that a custom FIR filter can be built with the standard FIR structure (a process to be discussed later), few design tasks remain other than the generation of the coefficients.

The coefficients for a FIR filter can be designed in many ways. We would need another whole book to describe all the methods. Instead, we're going to describe perhaps the simplest, most general way to design a FIR filter. The technique uses Fourier transforms and a technique called windowing. We won't go fully into exactly why this technique works, but rather how it works.

The technique is general because it enables the construction of a filter with an arbitrary frequency transfer function. The designer can describe a custom-shaped frequency

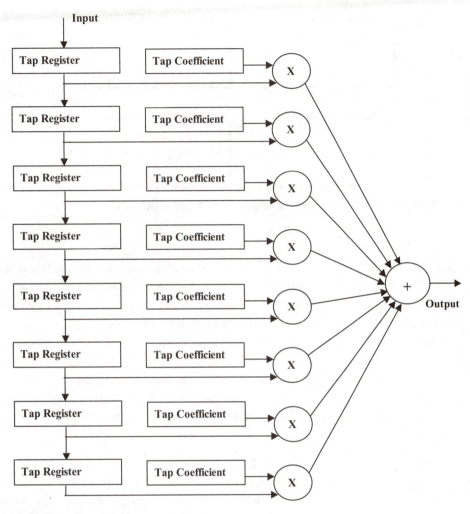

FIGURE 8-11   FIR filter structure

response (within bounds) and then apply the techniques. In practice, most filters have very specific functions and the following four filters are the most commonly used designs. Figure 8-12 shows low-pass, high-pass, band-pass, and band-stop filters:

■ **Low-pass**   The low-pass filter is designed to eliminate frequencies above the filter's cutoff frequency. Primarily, the cutoff frequency and the cutoff attenuation characterize the filter. It is commonly used to eliminate high-frequency noise or as an antialias filter.

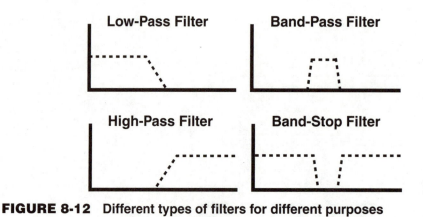

FIGURE 8-12 Different types of filters for different purposes

- **High-pass**   The high-pass filter is designed to eliminate frequencies below the filter's cutoff frequency. Primarily, the cutoff frequency and the cutoff attenuation characterize the filter. It is commonly used to eliminate a 60 Hz hum in systems or to accentuate high-frequency components in audio channels.
- **Band-pass**   The band-pass filter is designed to attenuate all frequencies except those within a narrow band. The filter is characterized primarily by the two frequencies (start of band and end of band) and the cutoff attenuation.
- **Band-stop**   The band-stop filter is designed to attenuate all frequencies within a narrow band. The filter is characterized primarily by the two frequencies (start of band and end of band) and the cutoff attenuation.

The Fourier approach to designing an FIR filter starts with the required shape of the filter transfer function. The four previous filters are examples, and we will move forward with the low-pass example. The math that follows is general and applies to any filter transfer function (within certain bounds). The URLs cited later allow designers to specify filter parameters and start a computation. The computations executed on the web sites use math similar to the math we'll describe next.

Subject to conditions, a simple filter's frequency response can be put in the general form:

$$F(j\omega) = \Sigma_{(n = 0, N - 1)}(c(n) \times e^{-jn\omega})$$

where N will become the number of taps in the FIR filter. c(n) will become the coefficient of the nth tap. Or by mathematical substitution,

$$F(j\omega) = \Sigma_{(n = 0, N - 1)}(c(n) \times (cos(n\omega) - j\,sin(\,n\omega)))$$

Figuring out the coefficient c(n) from this formula might involve some difficult calculus with an integral over a range of 2π. This is the case for a general-purpose (custom) frequency response, but if the frequency response curve is like the low-pass filter, the calculations are simpler. The gain is flat at a value of 1 and then drops off completely (in the ideal math equation). Taking advantage of the simplified filter shape, and with a few other mathematical manipulations, the integral reduces to a closed math solution as follows:

$$c(n) \;=\; (sin\,(n\omega)/n\pi)$$

Using the math identity sinc $(x) = \sin(x)/x$,

$$c(n) \;=\; \omega\,sinc\,(n\omega)/\pi$$

The sinc function is well known as the spectral envelope of a train of pulses. Figure 8-13 shows the shape of the sinc function.

One of the difficulties of the Fourier method is that it produces an infinite set of coefficients. This presents a problem because we cannot have an infinite number of taps in the FIR filter. If we simply eliminate some taps, the filter won't work as designed or simulated.

Instead, various techniques are used to minimize the taps to a conveniently small number. These techniques create a window value for every coefficient in the infinite series. All the coefficients are multiplied by the window during the FIR filter computations. All these windows limit the number of coefficients to the desired number of taps because the window has a value of zero for taps outside the range of the window.

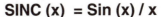

SINC (x) = Sin (x) / x

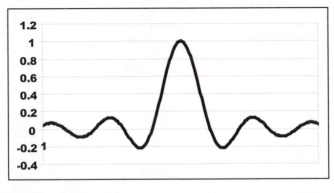

FIGURE 8-13 The sinc function

This means the FIR filter can be limited to a specific number of taps based on the window. Most of these windows keep the center taps (generally with the largest coefficients) and decrease the size of the window to zero as it reaches the edge coefficients.

The windows have well-known names and predictable effects on the filter. They are automatically added to the calculations since a window must be used to have a calculation at all. The URLs that follow allow us to perform calculations using JAVA tools. They have the windows built in to the Java tool that computes the coefficients and shows you the resultant filter transfer function. Each window has its strength and weaknesses, but we must choose a window for every calculation. Some of the windows are outlined here. In each case, we show the shape of the window. In addition, we show a FIR filter built with all the same parameters except for the choice of window type.

- **Rectangular window**   The rectangular window simply sets every window value to 1 around the center coefficient. This is true right to the edge of the filter. Outside the filter, all the coefficients are zeroed out of the window. The window chart has a characteristic rectangular shape. The rectangular window is easy to compute on the fly since only multiplication by unity is required. Most FIR filter coefficients, however, are precomputed during the design phase (see Figure 8-14). The math behind the rectangular window is explained at http://mathworld. wolfram.com/UniformApodizationFunction.html.
- **Bartlett (triangular) window**   The triangular window simply sets every window value to a linearly decreasing value starting at the center coefficient. Right at the edge of the filter, it reaches zero. Outside the filter, all the coefficients are

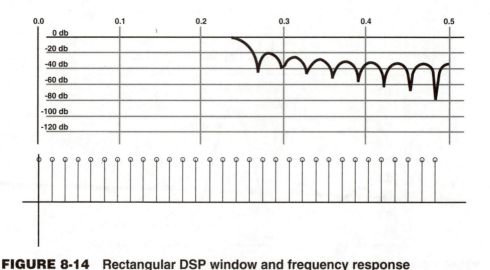

FIGURE 8-14   Rectangular DSP window and frequency response

zeroed out of the window. The window chart has a characteristic triangular shape (see Figure 8-15). The math behind the Bartlett function is explained at http://mathworld.wolfram.com/BartlettFunction.html.

■ **Hanning window**   This window is used to implement the Raised Cosine filter that we'll discuss later (see Figure 8-16). The math behind the Hanning window is shown at http://mathworld.wolfram.com/HanningFunction.html.

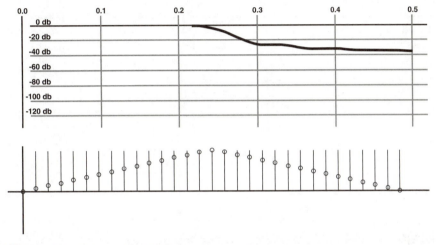

FIGURE 8-15   Triangular DSP window and frequency response

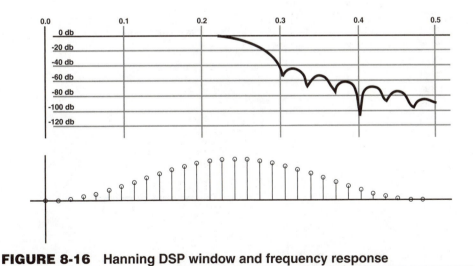

FIGURE 8-16   Hanning DSP window and frequency response

- **Hamming window**   This is a minor modification of the Hanning window (see Figure 8-17). The math behind the Hamming window is shown at http://mathworld.wolfram.com/HammingFunction.html.
- **Blackman window**   Similar to the Hamming and Hanning windows, the Blackman window has an extra term to reduce the ripple (see Figure 8-18). The

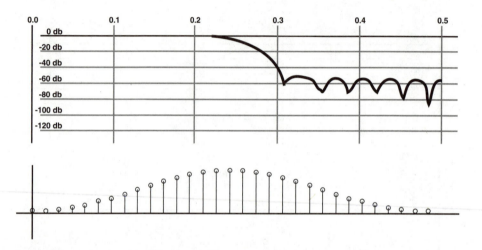

FIGURE 8-17   Hamming DSP window and frequency response

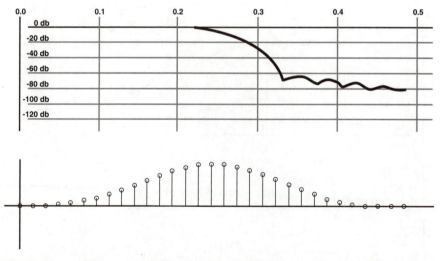

FIGURE 8-18   Blackman DSP window and frequency response

math behind the Blackman window is shown at http://mathworld.wolfram.com/ BlackmanFunction.html. More windows are shown at these sites:

- http://astronomy.swin.edu.au/~pbourke/analysis/windows/
- http://mathworld.wolfram.com/ApodizationFunction.html
- www.filter-solutions.com/FIR.html#asinxx

Among the web sites dedicated to filtering, the FIR Filter Design by Windowing site has a nice user interface where you can see the results of an FIR filter design (http://web.mit.edu/6.555/www/fir.html). It was used to make this chapter's figures. To use the tool, change the parameters, reselect the window type on the top pulldown list to recompute the coefficients, and redisplay the results.

In playing with this utility, I suggest altering just one parameter at a time. Try running a few other experiments as well. Notice how increasing the number of taps makes the filter rolloff sharper. Also notice that the ripple in the filter is largely unaffected by having more taps.

Physical Implementation of DSP Filters

As we mentioned before, all the DSP techniques we've mentioned so far are mathematical in nature.

FIR FILTERS

The physical implementation of antialiasing and dithering circuits notwithstanding, the structure of a FIR filter is theoretical: a series of registers, coefficients, and adders that form an arithmetic output. The DSP calculations can be performed in hardware or software. In most cases, the calculations could be done either way.

Software

Those of us who build hardware for a living can relate to feelings of frustration when it comes to DSP software. Somehow DSP programmers feel the DSP answers just float out of the air, computations unsullied by the presence of hardware or electrons. The truth is, DSP computers are very much hardcore hardware, specially designed for DSP calculations. We've discussed DSP computers previously in the book, so I won't go into the structure. The DSP chips are specially designed to be efficient at handling the types of calculations that are required for FIR filters. Specific logical structures within the

DSP can be used as a string of FIR registers and coefficient registers. Also, structures are used to move data efficiently through the DSP chip as rapidly as possible. DSP programmers can take advantage of many library functions. Implementing a simple FIR filter can be accomplished just by specifying the number of taps and the coefficients. The DSP compiler takes care of the rest of the work.

Hardware

Well, enough ranting about software and hardware people. The sad truth is, we need each other. Even the pure hardware implementation of FIR filters requires a significant amount of software tools and programming. Prepackaged implementations of FIR filters are available, but not common. The most common way they are implemented is in *Application-Specific Integrated Circuits* (ASICs) or FPGAs. FPGAs contain many registers and logic elements that can be configured using software. The software is typically written in higher-level languages like VHDL or Verilog. The VHDL code lines engender tap registers, coefficient registers, and *Multiply and Accumulate* (MACs). The entire FIR filter structure is visible right in the code itself. When the VHDL code is compiled and loaded into an FPGA, the FIR filter takes on a physical instantiation.

Here are some web sites describing FIR filter design in such languages:

- www.doulos.com/fi/vhdl_models/model_9605.html
- www.item.uni-bremen.de/research/papers/paper.pdf/Helge.Bochnik/nato93/boc9301.pdf
- www.altera.com/support/examples/verilog/ver_base_fir.html

Testing FIR Filters

Several easy tests can be run on a FIR filter design when it is first tested. Some tests are so simple they can be built right into the physical implementation. This allows the test to be executed at a later time. The FIR filter tests are as follows:

- **Coefficient test**   Feed the FIR filter a series of data points consisting of all zeroes with a single full value in the middle of the stream. As the full value hits each FIR filter tap along the way, the output will be a serial stream equal to all the coefficients right in order.
- **Frequency sweep**   To test any filter, analog or DSP, sweep it with a series of pure sine waves. The frequency response curve should be similar to that shown in the DSP design software. Further, if we continue the sine wave sweep above the Nyquist frequency, we should observe the effects of the antialias filter. If we

observe a significant response from the filter above the sampling frequency, we should reexamine the integrity of the antialias filter design. The output sine waves should be clean and well behaved.

The FIR Filter FAQ site contains a thorough explanation of FIR filters and lists a few more tests that can be run (www.dspguru.com/info/faqs/firfaq.htm). The following sites describe FIR filters and have various tools for designing them:

- http://web.mit.edu/6.555/www/fir.html
- www.nauticom.net/www/jdtaft/fir.htm
- www.filter-solutions.com/FIR.html#asinxx

INFINITE IMPULSE RESPONSE (IIR) FILTERS

Okay, now that we've wrestled FIR filters to the ground, here's another wrinkle. *Infinite Impulse Response* (IIR) filters are another option for designing a DSP filter. Although a FIR filter passes signals once through in a fixed, linear sequence, IIR filters have feedback loops. Output signals, even intermediate signals, are fed backwards during the processing. This has a few implications:

- *IIR filters are shorter.* Think for a minute about the path that data takes through an IIR filter. Instead of going through once, like in a FIR filter, the data may be fed back a few times. These extra loops through the IIR filters act almost as extensions of the filter. The result is that an IIR filter can get similar results with much fewer taps. Let's look at a rough comparison.

  Figure 8-19 is from a rectangular windowed FIR filter with 34 taps. It drops off 20 db in a frequency range of about 0.050 normalized.

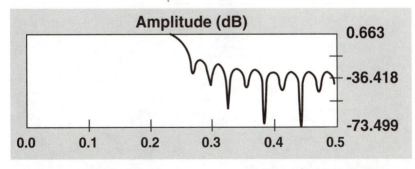

FIGURE 8-19   DSP FIR filter frequency response with a 34-tap filter

Figure 8-20 is from a twelfth-order Butterworth IIR filter. It too drops about 20 db in a frequency range of about 0.050 normalized.

But the IIR filter is just twelfth order, made out of a series of second-order IIR filters. A second-order filter can take many different structures. One example is shown in Figure 8-21. Each order is the hardware equivalent of about 2 FIR taps, so a twelfth-order IIR filter is the equivalent of about 24 FIR taps, shorter for the same results.

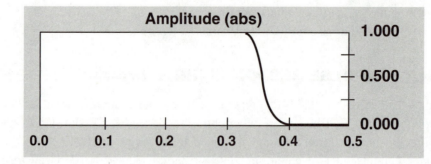

FIGURE 8-20 DSP FIR filter frequency response with a twelfth-order filter

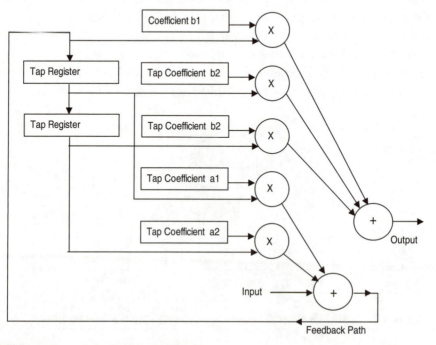

FIGURE 8-21 A second-order IIR filter

Diagrams for the design of IIR second-order filters can be found at http://spuc.sourceforge.net/iir_2nd.html and at www.nauticom.net/www/jdtaft/ biquad_section.htm.

■ *IIR filters have phase shift.* The group delay of the FIR and IIR filters we just compared is shown in Figure 8-22 and Figure 8-23. The FIR filter has a relatively fixed delay of 16.5 periods, which might be expected for a 34-stage FIR filter sampled at twice the frequency. I suspect the chart should have shown a flat delay of exactly 17 periods. This means there will be a fixed but constant delay in the FIR filter output.

The IIR filter has a variable delay, depending on the frequency of the input signal. Slower signals have a zero delay! The IIR second-order stage has a straight-through path, so signals get through right off the bat. Higher-frequency signals have an increasing delay approaching 19 clock periods. Because most IIR filters have different delays at different frequencies, they generally distort signals in ways that FIR filters do not. This may be a small price to pay for the smaller real estate used up in the construction of an IIR filter (see Figure 8-23). Another web site about IIR filters can be found at www.dspguru.com/info/faqs/iirfaq.htm.

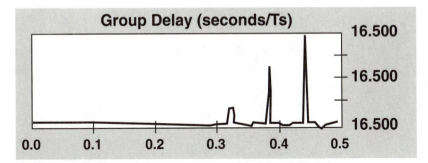

FIGURE 8-22 FIR filter delay

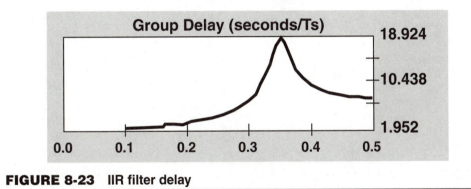

FIGURE 8-23 IIR filter delay

Multirate DSP

Multirate DSP filters are very similar to FIR and IIR filters, except data comes out of the filter at a different rate than it goes into the filter. We will not go into the exact techniques, but it bears mentioning in the book. This is used when sampled data is already available, but the data rate does not match the rate needed in a specific application. A specific example might be a digital video signal coming in at a full broadcast rate. At 270 million bits per second, it's might be too much data to send out over the Internet!

So the question is, how do we chop the data down to a lower bit rate even before we use MPEG to compress it for Internet transmission? It might make sense to decrease the video rate by a factor of three or five before sending it into the MPEG compression engine. A multirate DSP filter is perfect for this task. CommDesign offers a tutorial describing the basic techniques of multirate DSP at www.commsdesign.com/design_center/broadband/design_corner/OEG20020222S0071.

The following URLs have further information that might be useful in studying DSP:

- http://dspguru.com/info/tutor/index.htm
- http://ece-www.colorado.edu/~ecen4002/4 _filter_structures.ppt
- www.nauticom.net/www/jdtaft/
- www.dspguru.com/info/tutor/other.htm

Digital Signal Processing is a powerful tool we can use in the design of robots. If we pay attention to a few basic theorems and construct the DSP engine the right way, we can get very predictable performance.

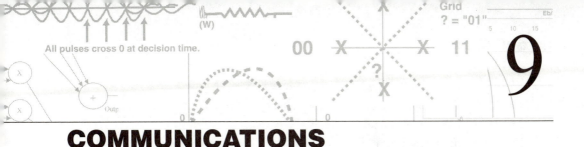

COMMUNICATIONS

It's not often one stares in the mirror and sees a perfect reflection, especially one that goes backward in time. But these things happen and they are not to be missed.

Take five minutes ago, for instance. I sat down in a quiet moment to reflect on how to teach the vast field of communications in one chapter. This is what I saw.

I spent eight years in English classes and not one of my teachers managed to convey to me the central purpose of their course. They were there to teach me how to communicate, from person to person. Such communication might happen through interactive conversation, through my writings, or through books. But not one of those eight teachers saw to it that I understood the basic purpose of the course. They failed to communicate, to me, the single most important piece of information they had to offer! Being a responsible adult, I do take responsibility for this. But what does this also say about our education system? I won awards for my achievements in English classes. And all the while even I knew that my English was crumby (sic)!

So I sat down and searched the entire Internet for the definition of communication. These were the URLs that turned up, in the very order that I searched them. This is what I found:

- WorldCom, a large communications company
  www.worldcom.com/global/resources/glossary/?attribute=term&typeOfSearch=
  2&searchterm=communications
  Defines communication as "The transmission or reception of information, signals,
  or messages.

- Merriam-Webster's, online dictionary
  www.m-w.com/cgi-bin/dictionary
  A process by which information is exchanged between individuals through a common system of symbols, signs, or behavior.

- St. John's Episcopal Church
  www.stjohnsdetroit.org/html-stj/06152000newsletter.html
  Offers that it is "The act of imparting or transmitting ideas, information, etc.

- Professor Robert J. Schihl
  www.regent.edu/acad/schcom/phd/com707/def_com.html
  Communication is a process in which a person, through the use of signs (natural, universal)/symbols (by human convention), verbally and/or non verbally, consciously or not consciously but intentionally, conveys meaning to another in order to affect change.

- Ted Slater
  www.ijot.com/ted/papers/communication.html
  Has this to say: "'Communication,' which is etymologically related to both 'communion' and 'community,' comes from the Latin communicare, which means, 'to make common' (Weekley, 1967, p. 338), or 'to share.' DeVito (1986) expanded on this, writing that communication is '[t]he process or act of transmitting a message from a sender to a receiver, through a channel and with the interference of noise' (p. 61). Some would elaborate on this definition, saying that the message transmission is intentional and conveys meaning in order to bring about change."

Okay, we can stop right here. Honest, these last two sites turned up in my random search. I'm going with Ted Slater, who probably spent some valuable hours with Professor Schihl. So today, kudos go to Regent University for not only stating a very clean definition of communication, but for broadcasting it to the world in a successful manner.

Readers wanting an alternate interpretation of Ted's web page are urged, again, to read R.D. Laing's book *The Politics of Experience*. Is it odd that it should take psychologists and professors at denominational universities to set the record straight?

So now I stand here with one chance to define what communication is. Here we go:

Communication is the process of sending information from source to destination.

Whoa. Don't jump yet. Here are my disclaimers.

- Nothing in my definition says the information has to arrive error free. Most information is sent with the full knowledge that it will be corrupted some en route. TV transmissions are surely in this category.
- Nothing in my definition says information cannot also go the other way during the same communication process. As long as information still gets from the source to the destination, the definition holds.
- I disagree that we must always ascribe motivation to the sender. Professor Schihl must argue his positions with passion! Although some communication is certainly useful in effecting societal change, much human communication is routine.
- The source and destination can be humans or machines. For that matter, some information is just sent to the dump, which hardly qualifies as communication. This makes the good professor's definition look a bit better!
- Most communication (99.9 percent?) falls on deaf ears. We need only go to the newspaper recycling plants to see this. Humans these days must be adept at tuning out the flood of communications coming at them from TV, radio, email, the Internet, and newspapers.
- Ted's expanded definition includes the communication channel and noise. These considerations are one layer down inside my definition. We'll get to them shortly.

So why is communications a topic in a book about robots? Well, we've entered an era where communication traffic is growing rapidly. Further, the amount of data stored in computers and data banks is growing rapidly as well. It's increasing something like 50 percent a year if we believe the storage industry hype.

Just as communication is vital to the effectiveness and power of people, so too will it become more important to robots. The modern employee is much more effective with the ability to get email and surf the Internet. As robots become more capable, communications will become more important to their design. At the very least, communication permits the remote monitoring of robots for many different purposes. To design robots well, a robot designer should have a firm grasp of communications.

Now, given that this is the twenty-first century, we are going to confine our discussion to digital communications and forgo all discussion of analog communications. True enough, digital communications do use analog electronics, but the prevailing mode of electronic communications today is digital. Cable TV, telephones, cell phones, and the Internet are all digital communications.

OSI Seven-Layer Model

Some years ago, a group got together in an attempt to define a model for the way communications should be structured, which was known as the *Open Systems Interconnection* (OSI) seven-layer model (www.scit.wlv.ac.uk/~jphb/comms/std.7layer .html). Nobody really followed the model from top to bottom, but *Transmission Control Protocol/Internet Protocol* (TCP/IP) network communication comes the closest; however, the model is useful at the very least as a checklist for the types of things we might want in a communications system. Given that it's also worth learning just for network communications, let's delve into it.

LAYER 1: PHYSICAL LAYER

The data layer is the lowest layer and defines the physical and electrical characteristics. It is the layer dealing with sending bits over the physical medium. All communications have a physical layer of some sort. In some systems, it may be the only layer. Baseband communications, modulation, demodulation, and transmission through the channels are all topics that loosely belong in this layer.

LAYER 2: DATA LINK LAYER

This layer deals with blocks of data on the physical media. It controls the sharing of the communication path, frames, flow control, and some low-level error checking. This is the *multiple access* (MAC) layer in network communications. Many strategies exist for sharing access to a transmission channel. Access and error-checking techniques are topics we can cover that belong to this layer.

LAYER 3: NETWORK LAYER

This layer is responsible for routing, making, maintaining, and breaking connections. This is the IP layer in network communications.

LAYER 4: TRANSPORT LAYER

This layer is responsible for the error-free transmission of data from one machine to another. This is the TCP layer in network communications.

LAYER 5: SESSION LAYER

This layer handles the life of the current connection and keeps the data traffic moving.

LAYER 6: PRESENTATION LAYER

This layer handles the data from applications. It performs packing, encryption, decryption, compression, and so on.

LAYER 7: APPLICATION LAYER

This layer is where the application software resides. More information about the seven-layer model can be found at the following PDF and web sites:

- www.itp-journals.com/nasample/t04124.pdf
- www.itp-journals.com/OSI_7_layer_model_page1.htm
- www.scit.wlv.ac.uk/~jphb/comms/std.7layer.html
- www.cs.cf.ac.uk/Dave/Internet/node51.html

Not everyone is happy with the seven-layer OSI model. Check out www.randywanker .com/OSI/ (rated R) and www.scit.wlv.ac.uk/~jphb/comms/osirm.crit.html

A couple of underlying ideas are behind the layering of this stack that applies across most communications:

- **Hidden functions**   The stack layers interact with a fixed interface. Portions of the stack can be redesigned internally and still function properly.
- **Common interfaces**   Because the stack layers interact with a fixed interface, two *different* machines can communicate with each other without a problem. They simply communicate from the same level to the same level. For example, TCP information at level 4 in one machine travels down the stack to the physical level and is sent to the other machine. At the receiving machine, it enters the physical level and travels up to level 4 where it appears as TCP information again.

Many communication techniques lead to standards that can be observed by all designers at various stack levels. Most communication standards are limited to just a few levels of complexity. They all have physical and link layers. Many have network and transport levels, but not many go to higher levels.

Physical Layer

All that said, digital communication comes down to one thing: sending data over a channel. Another fundamental theorem came out of Shannon's work (first mentioned in Chapter 8). It comes down to an equation that is the fundamental, limiting case for the transmission of data through a channel:

$$C = B \times log_2 (1 + S/N)$$

C is the capacity of the channel in bits per second, B is the bandwidth of the channel in cycles per second, and S/N is the signal-to-noise ratio in the channel.

Intuitively, this says that if the S/N ratio is 1 (the signal is the same size as the noise), we can put almost 1 bit per sine wave through the channel. This is just about baseband signaling, which we'll discuss shortly. If the channel has low enough noise and supports an S/N ratio of about 3, then we can put almost 2 bits per sine wave through the channel.

The truth is, Shannon's capacity limit has been difficult for engineers to even approach. Until lately, much of the available bandwidth in communication channels has been wasted. It is only in the last couple of years that engineers have come up with methods of packing data into sine waves tight enough to approach Shannon's limit. Shannon's Capacity Theorem plots out to the curve in Figure 9-1.

There is a S/N limit below which there canot be error free transmission. C is the capacity of the channel in bits per second, B is the bandwidth of the channel in cycles

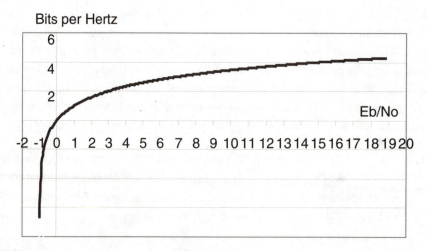

FIGURE 9-1 Shannon's capacity limit

per second, S is the average signal power, N is the average noise power, No is the noise power density in the channel, and Eb is the energy per bit. Here's how we determine the S/N limit:

$$S/C = Eb$$
$$N = No \times B$$
$$C = B \times log_2(1 + S/N)$$
$$C/B = log_2(1 + S/(No \times B))$$

Since

$$S = Eb \times C$$
$$C/B = log_2(1 + (Eb \times C)/(No \times B))$$

Raising to the power of 2,

$$2^{C/B} = 1 + (Eb \times C)/(No \times B)$$
$$Eb/No = (B/C) \times (2^{C/B} - 1)$$
$$Eb \times C/No \times B = 2^{C/B} - 1$$

If we make the substitution of the variable x = Eb × C/No × B, we can use a mathematical identity. The limit (as x goes to 0) of $(x + 1)^{1/x} = e$.

We want the lower limit of capacity as the S/N goes down. In the limit, x goes to zero as this happens. We have to transform the last equation and take the limit as x goes to zero.

$$Eb \times C/No \times B = 2^{C/B} - 1$$
$$1 + Eb \times C/No \times B = 2^{C/B}$$
$$log_2(x + 1) = C/B$$
$$x \times log_2(x + 1)^{1/x} = C/B$$
$$log_2(x + 1)^{1/x} = No/Eb$$
$$limit\ No/Eb = log_2 e = 1.44$$
$$limit\ Eb/No = .69$$

In dB, this number is -1.59 dB. Basically, if the signal is below the noise by a small margin, we are toast! Figure 9-1 shows this limit on the leftside.

This sets the theoretical limit that any modulation system cannot go beyond. It has been the target for system designers since it was discovered. The limit will show up below in the error rate curves of various modulation schemes.

Many ways exist for jamming electrons down wires or waves across the airways. In all these cases, the channel has a bandwidth. Sometimes the bandwidth is limited by physics; sometimes the *Federal Communications Commission* (FCC) limits it. In both cases, Shannon's Capacity Theorem applies: putting God and the FCC on equal mathematical footing.

A quick aside about the FCC: After college, we constructed and ran a pirate radio station out of a private house. We broadcast as WRFI for about two years, playing the music we felt like playing and rebroadcasting the BBC as our newscast. I was a DJ and a peripheral player. We had fake airwave names to hide our identities; mine was Judge Crater. Finally, after a great run, the FCC showed up at our door to shut us down. They had tracked us down in a specially modified station wagon with a directional antenna molded into the roof. They only had to follow a big dashboard display arrow to our door. It turns out the DJ at the time was playing a Chicago blues album. The FCC agents confessed that they liked the music so much that they pulled over until the album was complete before they knocked on the door. The DJ opened the door, the FCC employee folded open his wallet just like Jack Webb on Dragnet, and the DJ got a look at the laminated FCC business card. Both sides, in turn, dissolved in laughter. Two hours, and some refreshments later, they departed with our crystal, a very civilized conflict. But I digress.

Here are a couple of web sites and a PDF on Shannon's Capacity Theorem:

- www.owlnet.rice.edu/~engi202/capacity.html
- www.cs.ncl.ac.uk/old/modules/1996-97/csc210/shannon.html
- www.elec.mq.edu.au/~cl/files_pdf/elec321/lect_capacity.pdf

Every method of sending data across a channel has a mathematical footing. Often, the method itself leads to a closed mathematical form for the capacity of the method. Once the method is implemented, then the implementation can be tested using Shannon's Capacity Theorem. Calibrated levels of noise can be added to a perfect channel and the data-carrying capability can be measured. The testing methods are very complex and are shown at www.elec.mq.edu.au/~cl/files_pdf/elec321/lab_ber.pdf.

Baseband Transmission

Given a wire, it's entirely possible to turn the voltage off and on to form pulses on the wire. In its crudest form, this is baseband transmission, a method of communication distinct from modulated transmission, which we'll discuss later.

Baseband transmission is used with many different types of media. Data transmission by wire has occurred since well before Napoleon's army used the fax machine. Yes, the first faxes dropped on the office floor about that time in history (www .ideafinder.com/history/inventions/story051.htm).

Baseband transmission is also used in tape drives and disks. Data is recorded as pulses on tape and is read back at a later time.

A sequence of pulses can be constructed in many different ways. Engineers have naturally come up with dozens of different ways these pulses can be interpreted. As is often the case, other goals exist besides just sending as many bits per second across the channel as possible. However, in satisfying other goals, channel capacity is sacrificed. Here's a list of other goals engineers often have to solve while designing the way pulses are put into a channel:

- **Direct Current (DC) balance** Sometimes the channel cannot transmit a DC voltage at all. A continuous string of all ones might simply look like a continuously high voltage. Take, for instance, a tape drive. The basic equation for voltage and the inductance of the tape head coil is

$$V = L \times dI/dt$$

  V is the input signal, L is the inductance of the tape head's coil, and I is the current through the coil. If V were constant, we'd need an ever-increasing current through the coil to make the equations work. Since this is impossible, tape designers need an alternate scheme. They have come up with a coding of the pulses such that an equal number of zeroes and ones feed into the tape head coil. In this way, the DC balance is maintained. Only half as many bits can be written as before, but things work out well. The codes they use are a version of *nonreturn to zero* (NRZ).

- **Coding for cheap decoders** Some data is encoded in such a way that the decoder can be very inexpensive. Consider, for the moment, pulse-width-encoded analog signals. A pulse is sent every clock period, and the duty cycle of the pulse is proportional to a specific analog voltage. The higher the voltage, the larger the duty cycle, and the bigger percentage of time the pulse spends at a high voltage. At the receiver, the analog voltage can be recovered using just a low-pass filter consisting of a resistor and a capacitor. It filters out the AC values in the waveform and retains the DC. These types of cheap receiver codes are best used in situations where there have to be many inexpensive receivers.

- **Self-clocking** Some transmission situations require the clock to be recovered at the receiving end. If that's the case, select a pulse-coding scheme that has the clock built into the waveform.

- **Data density** Some pulse-coding schemes pack more bits into the transmission channel than others.

■ **Robustness**   Some pulse-coding schemes have built-in mechanisms for avoiding and/or detecting errors.

The following PDFs and web site provide a good summary of the advantages and disadvantages of various coding methods:

■ www.elec.mq.edu.au/~cl/files_pdf/elec321/lect_lc.pdf
■ http://murray.newcastle.edu.au/users/staff/jkhan/lec08.pdf
■ www.cise.ufl.edu/~nemo/cen4500/coding.html

PULSE DISTORTION: MATCHING FILTERS

One of the difficult problems with the transmission of pulses through a channel (wire, fiber optics, or free space) is that the pulses become distorted. What actually happens is that the pulses spread out in time. If the overall transmission channel has sharp frequency cutoffs, as is appropriate for a densely packed channel, then the pulses come out of the receiver looking like the sinc function we looked at earlier. The pulse has spread out over time (see Figure 9-2).

If we try to pack pulses like this tightly together in time, they will tend to interfere with each other. This is commonly called *Intersymbol Interference* (ISI), which we will discuss later (see Figure 9-3).

But there's a kicker here. A transmission channel cannot be perfect, with sharp rolloffs in frequency. As a practical matter, we must allow extra bandwidth and relax our requirements on the transmission channel and the transmission equipment. A common solution to this problem is the *Raised Cosine Filter* (RCF), a filter we saw before in Chapter 8 as the Hanning window. A common practice is to include this matching RCF in the transmitter to precompensate the pulses for the effect of the channel. The

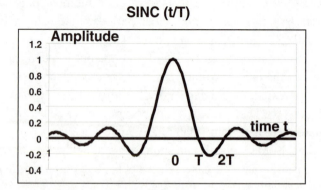

SINC (t/T)

FIGURE 9-2   Received pulses spread out to look like the sinc function.

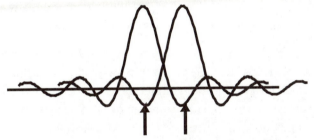

Intersymbol Interference

FIGURE 9-3   A poor receive filter enables consecutive pulses to interfere with each other.

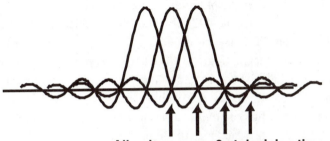

All pulses cross 0 at decision time.

FIGURE 9-4   A good raised cosine receive filter makes consecutive pulses cooperate.

received pulse signals, even though they have oscillations in their leading and trailing edge, cross zero just when the samples are taken. That way, adjacent pulses do not interfere with one another (see Figure 9-4).

The following sites discuss the RCF:

- www.iowegian.com/rcfilt.htm
- www-users.cs.york.ac.uk/~fisher/mkfilter/racos.html
- www.ittc.ukans.edu/~rvc/documents/rcdes.pdf
- www.nuhertz.com/filter/raised.html ·

COMMON BASEBAND COMMUNICATION STANDARDS

The following are some relatively common wired baseband communication links that we all have used. These are communication links that have relatively few wires and are generally considered serial links. Many computer boards come already wired with these sorts of communication ports, and many interface chips are available that support them.

■ **RS232/423**   RS232/423 has been around since 1962 and is capable of sending data at up to 100 Kbps (RS423) over a three-wire interface. It is considered to be a local interface for point-to-point communication. It's supposed to be simple to use, but it can cause a considerable amount of grief because many optional wires and different pinouts exist for various types of connectors. Other than the physical layer and the definition of bit ordering, very little layering takes place above the physical layer with RS232. For more info, go to www.arcelect.com/rs232.htm and www.camiresearch.com/Data_Com_Basics/RS232_standard.html.

■ **RS422**   RS422 uses differential, balanced signals, which are more immune from noise than RS232's single-sided wiring. Data rates are up to 10 Mbps at over 4,000 feet of wiring. Other than the physical layer and the definition of bit ordering, very little layering is done with RS422 (also see www.arcelect.com/rs422.htm).

■ **10BT/100BT/1000BT networking**   Ethernet is one of the most popular *local area network* (LAN) technologies. 10BT LAN technology enables most business offices to connect all the computers to the network. The computers can transmit data to one another at speeds approaching 9 to 10 million bits per second. As a practical matter, on busy networks, the best rates a user can achieve are much lower. The software stack includes up to four layers from physical layer 1 (*network interface* [NIC] cards), up to IP, and to TCP at layer 4.

100BT is 10 times faster than 10BT. 1000BT is 10 times faster again and available for use with a fiber-optic physical layer as well as copper wiring. See these web sites and PDF files for more info:

■ www.lantronix.com/learning/tutorials/
■ www.lothlorien.net/collections/computer/ethernet.html
■ ftp://ftp.iol.unh.edu/pub/gec/training/pcs.pdf
■ www.10gea.org/GEA1000BASET1197_rev-wp.pdf

Modulated Communications

Sometimes digital communications just cannot be sent over a channel without modulation; baseband communications will not work. This might be the case for several reasons:

■ Sometimes wiring is not a possibility because of distance. Unmodulated data signals are generally relatively low in frequency. Transmitting a slower baseband signal through an antenna requires an antenna roughly the size of the wavelength of the signal itself. For an RS232 signal at 100 Kbps, the signal has a waveform with about 10 microseconds per bit. Light travels 3,000 meters, about 2 miles, in 10 microseconds. We'd need an antenna two miles long to transmit such a signal efficiently into the impedance of space. Clearly, this won't work well. It's one of the primary reasons almost no baseband wireless communication systems exist. They almost all use modulation.

■ Sometimes the channel is so noisy that special techniques must be used to encode the signal prior to transmission.

■ The FCC and other organizations regulate the use of transmission spectra. Communication links must be sandwiched between other communication links in the legal communication bands. To keep these competing communication links separate, precision modulation is used.

Modulation generally involves the use of a carrier signal. The information signal (I) is mixed (multiplied by) the carrier signal (C), and the modulated signal (M) is broadcast through the communication channel:

$$M = I \times C$$

Although many different signals can be used as the carrier C, the type of signal most often used is the sine wave. Although the operation x can be just about any type of operation, the most common type of mixing involves multiplication.

A sine wave only has a few parameters in its equation. Thus, modulating a carrier sine wave can only involve a few different operations:

$$C = A \times sin (\omega \times t + \theta)$$

where A is the amplitude, ω is the frequency, and θ is the phase.

Any modulation of this carrier wave by the data must involve a modification of one or more of these three parameters. One or more of the parameters (A, ω, or θ) may take on one or more values based on the data. As the data input, I, takes on one of n different values, the modulated carrier wave takes on one of n different shapes to represent the data I. The following 3 discussions describe modulating A, ω, and θ in that order.

■ *Amplitude Shift Keying* (ASK) sets

$$M(n) = An \times sin (\omega \times t + \theta)$$

where A is one of n different amplitudes, ω is the fixed frequency, and θ is the fixed phase. In the simplest form, n = 2, and the waveform M looks like a sine wave that vanishes to zero whenever the data is zero (A = 0 or 1).

- *Frequency Shift Keying* (FSK) sets

$$M(n) = A \times sin(\omega n \times t + \theta)$$

where A is the fixed amplitude, ωn is one of n different frequencies, and θ is the fixed phase. In the simplest form, n = 2, and the waveform M looks like a sine wave that slows down in frequency whenever the data is zero (ω = freq0 or freq1).
- *Phase Shift Keying* (PSK) sets

$$M(n) = A \times sin(\omega \times t + \theta n)$$

where A is the fixed amplitude, ω is the fixed frequency, and θn is one of n different phases. In the simplest form, n equals 2, and the waveform M looks like a sine wave that inverts vertically whenever the data is zero (θ = 0 or 180 degrees).

Each modulation method has a corresponding demodulation method. Each modulation method also has a mathematical structure that shows the probability of making errors given a specific S/N ratio. We won't go into the math here since it involves both calculus and probability functions with Gaussian distributions. For further reading on this, please see the following web site and PDF file:

- www.sss-mag.com/ebn0.html
- www.elec.mq.edu.au/~cl/files_pdf/elec321/lect_ber.pdf

What comes out of the calculations are called Eb/No curves (pronounced "ebb no"). They look like the following figure, which shows a *bit error rate* (BER) versus an Eb/No curve for a specific modulation scheme (see Figure 9-5).

Remember, Eb/No is the ratio of the energy in a single bit to the energy density of the noise. A few observations about this graph:

- The better the S/N ratio (the higher the Eb/No), the lower the error rate (BER). It stands to reason that a better signal will work more effectively in the channel.
- The Shannon limit is shown as a box. The top of the box is formed at a BER of 0.50. Even a monkey can get a data bit right half the time! The vertical edge of the box is at an Eb/No of 0.69, the lower limit of the digital transmission we derived earlier. No meaningful transmission can take place with an Eb/No that low; the channel capacity falls to zero.
- This graph shows the BER we can expect in the face of various Eb/No values in the channel. Adjustments can be made. If the channel has a fixed No value that cannot be altered, an engineer can only try to increase Eb, perhaps by increasing the signal power pumped into the channel.

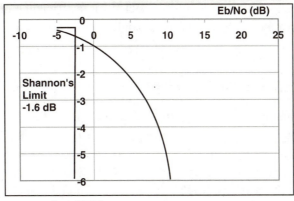

FIGURE 9-5 **S/N effect: As the power per bit (Eb/No) goes up, the bit error rate (BER) goes down.**

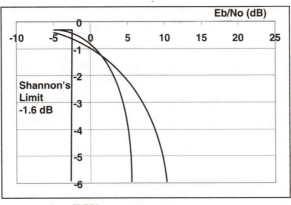

FIGURE 9-6 **A better modulator (the inner curve) can approach the Shannon limit more closely.**

■ Conversely, if an engineer needs a specific BER (or lower) to make a system work, this specifies the minimum Eb/No the channel must have. In practice, a perfect realization of the theoretical Eb/No curve cannot be realized and an engineer should condition the channel to an Eb/No higher than that theoretically required.

Figure 9-6 shows two BER curves from two different but similar modulation schemes. These curves show that some modulation schemes are more efficient than others. In fact, the entire game of building modulation schemes is an effort to try to approach the Shannon limit. As might be expected, more efficient modulators are more expensive. Most people settle for wasting bandwidth rather than paying for a more expensive modulator.

COMPLICATED MODULATORS

These previous examples are very rudimentary modulation schemes. Often, in modern modulation methods, more than one carrier parameter is modulated at the same time. Let's also introduce here the concept of a symbol. A symbol is simply a multiple bit number used for modulation. A byte could be an 8-bit symbol used in ASK to set the amplitude to one of 256 different levels. The process of modulating the carrier by a symbol changes the character of the carrier waveforms.

The receiver demodulates the data and makes an attempt to determine the character of the waveform in order to classify which symbol it represents. The demodulator in the receiver serves to quantify the received waveform into a symbol space. Visualize the symbol space as a multidimensional data space within which the received signal is moving. As the amplitude, frequency, and phase of the received signal change, the signal moves around in the receiver's symbol space. If, for instance, 256 different symbols are defined, then 256 different points are in the symbol space where these symbols reside. If the received signal is crossing one of these 256 points when the data clock ticks, the received symbol associated with that point is chosen as the received symbol, and the data (8 bits) represented by that symbol is dumped into the receiver's output.

Let's look at a simplified example. Suppose we are modulating both amplitude and phase with one bit each. Four different symbols (00, 01, 10, and 11) would be used and the symbol space might look like Figure 9-7.

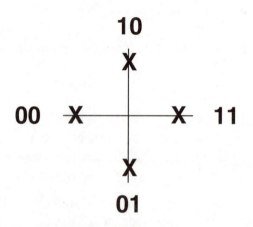

FIGURE 9-7 A graph of a simple symbol space

When the data clock ticks, we sample the position of the received signal in symbol space. Suppose we receive a symbol whose amplitude is a little low but has a very clear phase. It might map into the following point shown in Figure 9-8.

To decide on which symbol is received, we put a decision grid into symbol space, as shown in Figure 9-9. The decision grid makes the decision quickly, and the symbol is resolved to be 01.

It's clear that we do not want symbols to be too close together in symbol space. Modulation schemes are designed to minimize the probability that symbols will be too close or that the peculiarities of the channel will cause one symbol to be mistaken for another.

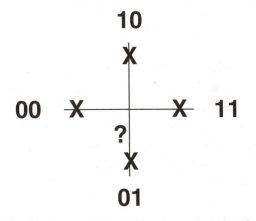

FIGURE 9-8 Classifying a recently received symbol that is shown as "?"

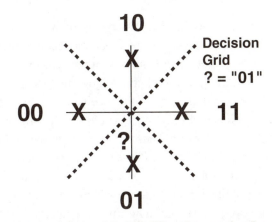

FIGURE 9-9 A hard decision grid classifies the received symbol as 01.

FIGURE 9-10 A symbol space for 64 QAM

A more complex example of this sort of symbol space is 64 *Quadrature Amplitude Modulation* (QAM), where 8 bits of data are modulated at the same time. Symbol space for 64 QAM might have a square structure as shown in Figure 9-10.

The incoming symbol data traces a wild pattern through the 8×8 grid of dots. To a certain extent, because the symbol data tries to stick to the grid points, the grid has open areas where the data does not traverse. These open areas look like eyes and are the subject of the next discussion.

Error Control

Designers of big symbol spaces have to worry about what's called the *open eye*. Remember, when the data clock ticks in the receiver, the received signal should be right on top of a symbol point. To get there from any other symbol point, it should travel along a well-known route through symbol space (governed by the shape of the carrier signal). With a noiseless channel, the trail of the received signal would trace a very nice set of geometric paths and lots of empty space would be showing on the symbol space, places where the signal never traverses. These empty spaces are what engineers look for when they are trying to find the open eye. These spaces are called that because they are generally formed by two sine waves and have the shape shown in Figure 9-11.

A good engineer can put the communication waveform on an oscilloscope (or other instrument), look at the eye pattern, and determine the health of the physical layer of the communication network.

Symbol 'X' is under each crossing

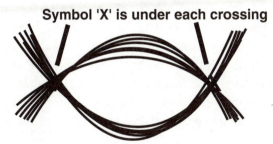

FIGURE 9-11   An open-eye diagram showing received signal traces crossing two symbol space X's

In the same manner, engineers can plot the recent data points to see how tightly they cluster around the symbol points. A healthy communication link will have a very tight clustering around the symbol points, and a sickly system will have them spread out in a sloppy manner.

These are all ways to try to keep the physical link healthy, but steps can be taken in the design of the communication link that will make it more robust. Many different ways are available for looking at what these techniques represent. I prefer to think of them all the same way: *sending the data more than once.*

In a situation where noise might ruin data inside the channel, the receiver is more likely to get the data if it's sent more than once. If the receiver is smart enough to recognize when data is corrupt, it can just wait for the second helping of the same data. This becomes particularly important for robots in remote locations.

Sending duplicate data can be done in many different ways. Clearly, it's possible to just send the data twice or three times. But believe it or not, it's possible to send the data 1.5 times, 1.1 times, or even 1.01 times.

Within certain bounds, robot designers can choose among communication protocol codes that enable them to pick the amount of redundancy built into the communication link. Since redundant data consumes bandwidth, this allows the designers to decide how much of the bandwidth is wasted. Sending extra data effectively lowers the BER, since errors are corrected at the receiver. Getting a lower BER is almost the equivalent of having a better Eb/No. Thus, designers can say they get coding gain out of different communication protocol codes. This coding gain can actually be realized since the coding gain can be subtracted off the Eb/No in the actual channel to get the same BER in a given situation. Add coding gain, decrease the Eb/No gain, and come out even. In practice, however, most engineers take the coding gain on top of the existing Eb/No and realize their profit as a lower BER.

This happens in satellite communications all the time. In fact, most satellite communication links are designed and specified with the coding gain built right into the communication protocol. Since many of the codes have parametric options, it is possible for the operator of a satcom link to pick a code on the fly that matches the quality of the channel. If the satcom link has a low No, then little coding gain may be needed and the data rate can go up. If the satellite link has a high No, then a stronger coding gain may be needed to maintain the quality of the data at the expense of a lower data rate.

ERROR DISTRIBUTION

Robot designers must also take a very careful look at the channel. It's one thing to predict the BER from the modulation method and coding, but it does no good at all if sunspots wreck the transmission for minutes or seconds at a time. Error rates are concatenated; all the links in the communication chain must be functioning at the same time. An error in any one link may, or may not, be corrected in another link down the chain.

In addition, noise is unpredictable. That's why they call it noise in the first place. Granted, it has certain mathematical properties that are dependable in the average, but random events can lead to a burst of errors that may not be caught by the coding scheme chosen. We must look at the density and distribution of errors in the channel, in addition to the error rate.

One thing further must be said about the distribution of errors. Some coding schemes (like Viterbi, which we'll get to soon) gather up errors all together in a net and correct them all at once. The problem is, if something goes wrong and they cannot all be corrected, the net rips and a local flood of errors happens that would not have occurred naturally in such a manner. This type of situation is actually *caused* by the error-correction coding scheme. The system must be prepared to survive such an event. We've probably all seen such error bursts in the middle of soccer games from overseas. The game goes along fine until there's a massive burst of black and green blocks on the screen. We'll see why this occurs shortly.

Let's take a look at some of the coding methods that send duplicate data. The different techniques have the same basic purpose: to decrease the error rate by sending some of the data more than once. The techniques are basically divided into two different methods. Some communication channels are bidirectional, and many are not. A bidirectional communication channel enables the retransmission of data by request of the receiver; a unidirectional communication channel does not.

BIDIRECTIONAL COMMUNICATION CHANNELS

A bidirectional communication channel enables the receiver to send the transmitter information about the state of the channel and the integrity of the received data. Several tools are used in a bidirectional communication channel to help send duplicate data. These tools are not confined to use in a bidirectional channel, but they can be used to take maximum advantage of the reverse communications link. In fact, all the tools used in a unidirectional communication channel will also work in a bidirectional channel.

BLOCK CHECKSUMS

When the receiver receives data, it must determine, to the extent possible, whether the channel has changed the data. It does not matter where in the channel the data was changed. Noise from lightening storms or sunspots may have changed the data en route or the receiver might have had a temporary power glitch. The only thing that counts is whether the receiver's data buffer got the same data that was transmitted. Much like aspirin bottles that come with a safety seal that ensures protection, data can be wrapped in a checksum that will guarantee the integrity of the data.

A checksum is a series of data bits that serve to summarize a block of data. The sender can chop the data stream into a series of blocks that may be many bytes long. The checksum is computed and appended to the data block before transmission. We'll discuss just how checksums are computed later. The receiver knows, by prior arrangement, how the checksum will be computed. The receiver, upon receiving the data block (and checksum), independently computes the checksum again and compares it to the received checksum. If the results are different, then a problem exists. If the checksums are the same, then the data is accepted and the receiver moves on to the next block. But suppose a problem exists. In this case, several different actions are possible.

Single Error Detection

If the transmitted checksum information has relatively few bytes, it's possible that an error can only be detected. There may not be enough information to either correct the error or to even detect more than one error in the data block. If an error is detected, the receiver can ask the transmitter to retransmit the block of information. One protocol used in the retransmission of data is discussed later.

Multiple Error Detection

If the checksum has enough data in it (and the appropriate mathematical structure), then it may be possible to detect more than one error in the data block. Note this means that a weak checksum method (with little data in the checksum) may even fail to detect any error if more than one error occurs in the data block.

Consider the nature of the communication channel used in the robot. If it is possible for more than one error to occur at the same time, then try a checksum method capable of at least detecting multiple errors. It is certainly possible for multiple errors to occur at the same time in any communications channel. The key question a robot designer should examine is the likelihood of such an occurrence. Examine the probability of errors and the distribution of the errors. Assuming the error rates are small and that the errors occur independently, it's safe to assume the chance of two simultaneous errors in a block is roughly the square of the chance of a single error in a block. The robot designer should compute this dual error rate and determine if it will be an acceptable error rate if such errors slip through.

Single Error Correction

If the checksum contains sufficient data to not only detect the existence of an error but correct it as well, then the data can be corrected before the receiver moves on to the next block of data. No retransmission from the transmitter will be required. It should be noted that even error correction schemes will occasionally make mistakes. The strength of the error-correcting code lies in the mathematics of the protocol. Some errors may not even be detected, some errors may not be correctable, and some errors will be incorrectly corrected. When employing such methods, the robot designer must examine these error rates and compare them to the allowable error rate.

Multiple Error Correction

Some checksums have sufficient information to correct simultaneous errors. All the same precautions should be taken as outlined previously. Be aware that such strong checksums often consume a good deal of bandwidth sending extra checksum data; the checksums may contain many bytes.

Checksums are smaller blocks of data that summarize larger blocks of data. Often checksums are called *cyclic redundancy checks* (CRC). The following web sites will point out a small difference. Certainly, if a checksum contains more data than the block it summarizes, then it is not of much use. The whole idea is to summarize the block of transmitted data in a small number of bytes in an effort to be efficient. Often, a checksum will consist of 1 to 8 bytes of extra information summarizing a block of data that is between 32 and 1,024 bytes long. These numbers are arbitrary, but common. TCP/IP, for instance, typically has blocks of data 512 bytes long with checksums that are 2 bytes long.

Descriptions of the IP checksum method can be found at:

- www.ietf.org/rfc/rfc1071.txt
- www.netfor2.com/checksum.html

Here are descriptions of TCP checksums:

- www.netfor2.com/tcpsum.htm
- http://ethereal.ntop.org/lists/ethereal-users/200012/msg00050.html

An interesting statistical analysis of TCP/IP checksum errors in a real-world application can be downloaded from www.acm.org/sigcomm/sigcomm2000/conf/paper/sigcomm2000-9-1.pdf.

The astute observer will note that a data block of 512 bytes can be filled in $2^{512 \times 8}$ different ways. However, a checksum with just 2 bytes can only take on 65,535 ($2^{2 \times 8}$) different checksum values. This means that for each possible checksum value, about 2^{256} (or about 7.4×10^{19}) data blocks will have the very same checksum.

So how do we get away with saying that this sort of checksum is sufficient for an application? If an error occurs, the erroneous data block just might be identical to one of the several billion data blocks with the same checksum. The key thing to remember is that a single error should result in an erroneous data block with only one chance in 65,536 of having the same checksum. If this decrease in the error rate is not good enough, then design the robot with a stronger checksum, which is perhaps longer. Certainly, as the mathematical algorithm is chosen for the checksum calculation, make sure the most common errors all result in a checksum change.

For example, an error in a single bit may be common and should result in a different checksum. The calculation method for checksums is often described by a polynomial, a mathematical way to describe the calculations involved in computing a checksum. The mathematics behind the selection of a good polynomial are beyond the scope of this book. Fortunately, many standard polynomials (some listed later) exist and we can select among them without reinventing them.

The following web sites describe using polynomials for the computation of checksums:

- www.4d.com/ACIDOC/CMU/CMU79909.HTM
- www.geocities.com/SiliconValley/Pines/6639/docs/crc.html
- www.relisoft.com/Science/CrcMath.html

- www.relisoft.com/Science/CrcNaive.html
- www.relisoft.com/Science/CrcOptim.html
- www.relisoft.com/Science/source/Crc.zip

PARITY BITS

Let's look at a simple checksum structure example. Parity bits, as part of a checksum structure, can simply indicate how many ones are in a byte. Basically, take a byte and count up the number of ones in the 8 bits. If we are using an even parity scheme, then the number of ones in the bits (including the parity bit) must be even. For example, if an even number of ones is in the data byte, then append a ninth parity bit containing a zero to the byte to keep an even parity. If the number of ones in the byte is odd, then append a one as the ninth parity bit to attain even parity. If we do this for every byte in the data block, then single bit errors in any byte will "finger" that byte as bad. We will be able to detect single bit errors in the data block at the expense of increasing the data by 1/8.

If we also compute the parity for each bit, over the entire data block we will get more capability. We can, for example, compute the number of ones in the 0 bit position for the entire data block and append a column parity byte at the end of the data block containing a single 9-bit number. The column parity byte will contain the parity computed for the 0th, first, second, . . . eighth, and ninth columns of bits in the data block. Then, if a single bit is corrupted in the data block, that byte's parity bit will signal which byte is erroneous, and the column parity byte will tell us which bit is wrong in that byte. This will allow us to correct single bit errors in a data block by duplicating and expanding the data block by about 1/8. It's not a very strong code; better ones can be created.

It is easy to make up our own code, but we must be sure it matches the requirements of the robot's operating environment. The strength of the code should match the error rates, the error distribution, and the tolerance the robot has for errors.

REED-SOLOMON CHECKSUMS

One of the most often used checksum calculations is the *Reed-Solomon* (RS) code. This type of code is capable of correcting multiple errors in a block of data. The reason this is useful will be outlined shortly. RS coding also expands the data block by appending parity bytes.

One popular RS code is RS(255,233), which expands a 233-byte data block to 256 bytes by appending 32 bytes of parity checksums, an expansion of the data block by a factor of about 14 percent. The RS(255,233) polynomial enables up to 16 different bytes to be corrected at the same time.

Another popular RS code is used in satellite video transmissions. The *Digital Video Broadcast-Satellite* (DVB-S) standard has been standardized on MPEG2 video transmission using, among other codes, RS(204,188). This code appends 16 parity checksum bytes to a data block of 188 bytes for a code expansion of about 8.5 percent. The RS(204,188) polynomial enables up to eight different bytes to be corrected at the same time.

The following web sites and PDF file outline RS encoding and decoding:

- www.4i2i.com/reed_solomon_codes.htm
- www.siam.org/siamnews/mtc/mtc193.htm
- http://web.usna.navy.mil/~wdj/reed-sol.htm
- http://reedsolomon.tripod.com/rs-encode.c
- www.elektrobit.co.uk/pdf/reedsolomon.pdf

For fun, go to www.mat.dtu.dk/people/T.Hoeholdt/DVD/index.html, which shows RS corrections in real time in a very graphic manner. The web page displays an image, shows graphically the amount of redundant data, enables us to introduce errors in the graphics image using the mouse, and corrects the errors before our eyes. If too many errors are introduced, the errors cannot be corrected. This illustrates the limits of block encoding.

RETRANSMISSION

If an error is detected, the receiver can send a NACK, or *Negative Acknowledge*, back to the transmitter. This NACK message will request the retransmission of the faulty data block. Some bidirectional communication protocols call for the receiver to transmit an *acknowledge* (ACK) message to acknowledge the reception of every perfectly good data block. If the communication channel imposes a significant delay on transmissions (such as what might occur to a remote space probe's robot), then sending an ACK (or NACK) message for every data block is impractical. If the transmission protocol enables the transmitter to transmit multiple blocks of data without receiving messages from the receiver, then the transmitter must append an identifier to each data block sent.

The identifier is often just a sequential count sufficient to distinguish each data block from its adjacent neighbors. The receiver, upon identifying a bad checksum, appends the identifier of the bad block to the NACK message for that block. When the transmitter receives the NACK message, it reassembles the data block that corresponds to the identifier and retransmits it. The receiver must compute the checksum of the received retransmission and accept the data block. Note that this will require both the receiver and the transmitter to buffer (keep) multiple blocks of data in memory during the transmission cycle.

Be aware that certain communication protocols cannot use retransmission as a tool to decrease errors. Video and audio links, for example, cannot use retransmission. Video and audio streams cannot pause while the data is retransmitted because the screen will go blank. These data streams must be continuously available at the transmitter and rely entirely on unidirectional data transmission (which we'll discuss shortly).

CHANNEL TUNING

A bidirectional communications link can be optimized in real time by sending control information in both directions. Channels can change over time and sometimes need tuning to work properly. Some communication protocols have built-in control signals and specified tuning algorithms that keep the communication link healthy and robust. The following methods can be used to tune a system:

- **Power**   A data communication link will often work better if more power is used to transmit each bit. The Eb/No ratio is directly affected. The receiver can measure the signal strength it is receiving from the transmitter. If it determines the signal is too weak, the receiver can send a request to the transmitter to boost its power when transmitting. In the same manner, the transmitter can request the receiver to boost its transmitting power. This technique can be used in all bidirectional communication links as long as the power stays within limits.

   What can be done with power control, however, is limited. Too much power can pollute the spectrum and make it impossible for any communication link to function properly. A properly constructed power control protocol for a communication link often includes a limit on the power that is received. If a receiver senses too much signal strength coming in from a transmitter, it can request the transmitter to decrease the signal strength to an acceptable level. After all, the signal for one receiver may just be the noise for another receiver. Some cooperation is therefore required.

   The *Code Division Multiple Access* (CDMA) protocol uses just such a power control protocol to optimize the communication link. This technique is especially useful in situations where a cellular phone is moving from one area to another in a car. The cellular base stations used by the phone change as the phone moves. To make sure the phone is well behaved and doesn't disturb the neighboring phones, power control is used. Here are some web sites and PDF files describing the technique further:

   - www.comsoc.org/livepubs/surveys/public/2000/dec/dukic.html
   - www.commsdesign.com/main/2000/09 /0009feat3.htm
   - http://vig.pearsoned.com/samplechapter/0130871125.pdf

■ **Code changes** If a communication link begins to deteriorate, another technique that can be used is a coding change. By prior agreement, the receiver and transmitter can pause and change coding methods. Stronger error correction codes translate directly to a coding gain that can be added to the Eb/No. As we discussed before, this generally means that an extra amount of redundant data will be sent in one form or another. Since extra data will be sent over the channel, and since the channel's Eb/No value is already marginal, it makes sense to move to a lower bandwidth for the data transmission. If less actual data is sent, more redundant data can be appended, and the channel power per bit remains the same.

A specific example of this can be found in MPEG video transmissions. Most MPEG transmissions are unidirectional, but some video links do have reverse control channels of a much lower bandwidth. Although video may be sent over unidirectional satellite links, the reverse control channel can be established over the phone.

At the transmitter site, an MPEG compressor takes a video signal and compresses it using the MPEG algorithms. The compressor has a choice of several compression algorithms that can squeeze the video picture down to smaller and smaller amounts of data (at the cost of picture quality). The compressor then encodes the MPEG data for transmission through the channel using Viterbi and RS codes that append redundant data. The receiver uses the Viterbi and RS codes to eliminate errors and then decompresses the video picture.

If the receiver cannot correct all the errors, the picture will begin to break up. The receiver can use the reverse control link to request a better channel coding method. The compressor at the transmitter site then uses a stronger compression algorithm to reduce the amount of data sent and chooses a stronger Viterbi and RS code combination. The channel coding increases the data back to the original amount again. The receiver will then be able to correct all the errors and present a clean picture. The video image may not be as good as before (because of the extra compression), but at least the images are going through.

UNIDIRECTIONAL COMMUNICATION CHANNELS

We've already discussed or mentioned many of the methods used to decrease errors in communication channels. Except for retransmission requests, which are impossible in a unidirectional communication channel, most of the same techniques can be used. We'll discuss a few more of the protocols used, but we won't go into great depth. However, to adequately specify a communications link for a robot, we must understand the options.

We need to realize that a unidirectional communications link can only be used successfully if the following two conditions are met:

■ The receiver's target error rate must be set so it is acceptable given the specifications for operation. We can pretty well determine ahead of time what error rate will be acceptable for operation of the robot.

■ The data received at the receiver must be of sufficient quantity and quality to keep the data rate high enough and the receiver's error rate below the acceptable target value.

To accomplish the second goal, we should review the tools available. In the case of bidirectional communications, we already talked about block encoding, channel tuning, and retransmission. Since both channel tuning and retransmission are impossible without a reverse communications channel, we should examine encoding further.

We've already discussed block encoding and checksums at some length. Parity bits and RS encoding are tools that can be used in a unidirectional communications link. Often, the name given to unidirectional error correction methods is *forward error correction* (FEC). It has this name because all error correction information moves forward; no reverse communication link exists. Here are a few sites about FEC:

■ www.its.bldrdoc.gov/fs-1037/dir-016/_2298.htm
■ http://research.compaq.com/SRC/articles/199711/error_correction.html
■ www.eccpage.com

Two other tools have proven valuable, namely convolution codes and concatenated codes.

CONCATENATED CODES

The general idea behind concatenated codes is to herd randomly spaced errors into one spot where we can dispatch them efficiently and reliably. That may be a gross oversimplification, but it is the way I view the technique (see Figure 9-12).

Figure 9-12 shows the typical arrangement for a communications system using concatenated codes. MPEG video signal data is broadcast in DVB format over satellites using this type of concatenated coding. We'll discuss MPEG compression and the DVB format later. The description of each block within the figure is as follows:

■ **MPEG compressor**   Broadcast video signals, generated by a video camera, are accepted by the input to the MPEG compressor. The compressor has several *digital signal processing* (DSP) computation engines that compress the signal. We will discuss data compression later.

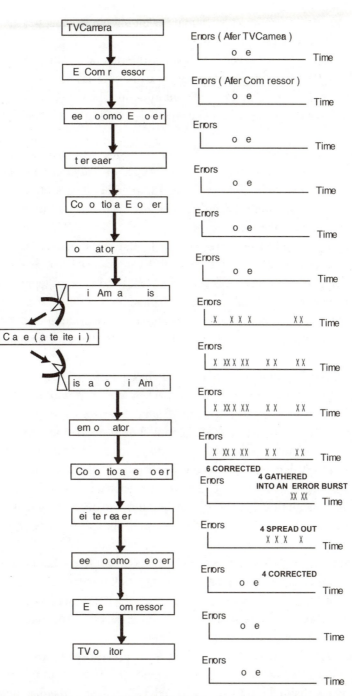

FIGURE 9-12 Satellite video broadcasting: concatenated coding showing the introduction and correction of errors

- **RS encoder** The compressed signal is sent into an RS encoder that adds checksum data as discussed previously.

- **Interleaver** An interleaver is a data shuffler that takes adjacent bytes and separates them. It does not expand the data block it receives, but it rearranges the order of the bytes in the data block. The goal of an interleaver is to arrange the data so the deinterleaver can separate adjacent errors, making them stand alone. We'll see how that works later.

- **Convolutional codes** The convolutional encoder effectively adds extra data to each data symbol. A couple of different types of convolutional codes exist, the most popular of which are Viterbi and Turbo codes. These codes tend to expand the data more than the RS encoding does, except the data is added almost byte by byte. We'll discuss these codes shortly.

- **Modulator** As discussed previously, the data modulator alters carrier waveforms according to the data transmitted. Even after the data is modulated once, the resulting waveform may be modulated a second time to step it up in frequency for specific communication frequency bands.

- **Channel** The data communications channel is taken to be a standard communications link (such as a satellite link) with errors added as the result of interference and noise.

- **Demodulator** A demodulator basically has the reverse function of a modulator. Often, the frequency will be stepped back down once with a first demodulator stage. The data will then be separated from the carrier wave in the final demodulation step. The demodulator data output should be identical to the modulator's data input, save for the errors introduced by the channel noise.

- **Convolutional decoder** The convolutional decoder effectively strips off the extra data the convolutional encoder added to each data symbol. The decoder must match the convolutional encoder. The output of the decoder should be identical to the input to the convolutional encoder, save for the errors introduced by the channel noise.

- **Deinterleaver** The deinterleaver is a data shuffler that takes adjacent bytes and separates them. It does not expand the data block it receives, but it rearranges the order of the bytes in the data block. The goal of a deinterleaver is to separate adjacent errors (bursts of errors) coming out of the decoder. This makes each bit error stand alone. We'll see how that works later.

- **RS decoder** The RS decoder, as discussed previously, strips off the checksum data and corrects errors as discussed previously. The output of the RS decoder, assuming all channel errors are corrected, is identical to the data output from the MPEG compressor.

- **MPEG decompressor** The decompressor has a DSP compute engine that decompresses the MPEG video data. The output of the decompressor is a broadcast video signal suitable for viewing.

Figure 9-12 shows the distribution of errors in an MPEG satellite transmission and helps explain why concatenated codes work so well. The figure shows the errors present in the DVB communications link. Errors are shown as tic marks on the time graphs to approximate the distribution over time. This shows the relative action of the various concatenated coding blocks. The following describes the action of each block in the figure with concentration on the handling of errors:

- MPEG compressor
- RS encoder
- Interleaver
- Convolutional codes
- Modulator

With luck and proper design, none of these five preceding blocks adds errors to the data. The modulator is the one block capable of introducing partial errors in the sense that it provides D/A functions. No analog signal is ever perfect. A good modulator will not add any significant errors.

Channel

The data communications channel is taken to be a standard communications link with errors added as the result of interference and noise. Data errors might occur at random intervals, or in concentrated bursts. Such errors are as follows:

- **Random errors** Random errors are the easiest to fix. The existing concatenated codes are well suited to fixing random errors.
- **Bursts of errors** The existing concatenated codes are reasonably well suited to fixing bursts of errors. The convolutional codes tend to concentrate errors into short bursts anyway. Naturally, if too many errors occur, they cannot all be corrected.
- **Regularly space errors** The existing concatenated codes have the most trouble with errors that occur at regular intervals. The RS block codes, in particular, are weakest at correcting such errors. This is not to say that these codes will not take care of errors distributed in such a manner. Just be careful designing a communication link if the noise is organized in some way.
- **Demodulator** By and large, a demodulator will not add much noise to the signals in the channel. It will add a small amount, but by the time a demodulator is

finished with its job, all the channel noise has been turned into digitized noise, data with some errors in it. The random noise from the channel is shown unchanged after demodulation.

■ **Convolutional decoder**   The convolutional decoder tends to gather errors together and correct them. When it fails, it lets through a burst of errors. By and large, it tends to gather errors together in time. This is a most useful function because it will make the output of the deinterleaver very predictable. The noise is shown gathered into bursts after the convolutional decoder.

■ **Deinterleaver**   The deinterleaver takes the bursts of errors that come out of the convolutional decoder and spreads them out evenly in time. This sets them all up individually to be picked off by the RS decoder and corrected. The noise is shown spread out into a regular pattern after the deinterleaver.

■ **RS decoder**   The RS decoder, as discussed previously, is capable of correcting multiple byte errors in a block of data. Most concatenated codes are constructed in such a way that the burst of errors coming out of the convolutional decoder does not, in general, contain more bytes than the RS decoder is capable of correcting. The noise is shown completely corrected after the RS decoder. This finished the stated goal of showing how the concatenated codes work. In practice, of course, some noise always gets through.

■ **MPEG decompressor**   The MPEG decompressor does not add any errors to the data stream.

DVB concatenated codes are covered further at www.csee.wvu.edu/~mvalenti/documents/milcom00.ppt. Such codes are used in the transmission of video data over satellite links.

CONVOLUTIONAL CODES

Convolutional codes increase the amount of data redundancy in a data stream. The decoders have memory within them and delay the output of data for a short while. Redundant data can be added in a couple of different ways.

Expanding the Bandwidth

Certainly, redundant data can simply be added to the existing data. If this is done, then extra bandwidth is required to transmit the extra data. We can illustrate this describing Viterbi codes.

Viterbi Encoder (in the Transmitter)   When we can expand the bandwidth, Viterbi codes specify a state machine that has the unexpanded data as an input. The state machine creates the output data. The state machine changes states depending on the flow of input data bits. Each one and zero from the unexpanded data does two things:

- **Changes states**   The state machine changes state for each one and zero coming in from the unexpanded data input.
- **Outputs data**   Each time the state machine receives an input bit and changes state, it outputs some data. In general, more bits are output than are input, so the Viterbi encoder expands the data.

Since more than just two states exist in the Viterbi state machine, each state change excludes some states in favor of the two states that are the only possible results of the change. If, for example, the state machine has four states, then a one or a zero can only make the state machine change to one of two different states. The other two states are prohibited changes. This is the key to how the Viterbi decoder corrects errors.

Viterbi Decoder   It's a great oversimplification, but here's a brief explanation of how the Viterbi decoder corrects errors. The Viterbi decoder knows a priori how the Viterbi encoder functions. Since the decoder knows the behavior of the encoder, it can detect suspicious data altered by the noise in the channel. The random changes caused by the noise cannot fully mimic the activity of the encoder. Thus, the decoder can detect signs of tampering.

The decoder accepts input bits coming directly out of the channel through the demodulator. The decoder also has a state machine that changes states based on the received data bits. As the expanded input data is received, the decoder state machine does two things:

- **Changes state**   As expanded data bits are received from the demodulator, they are decoded to determine the next state of the decoder state machine. The history of the state changes is accumulated back a few cycles.
- **Outputs data**   As the decoder state machine changes states, it triggers an examination of the historical state change data. If the historical data all makes sense, then the decoder outputs the unexpanded data that is stored in the historical data. In general, the output has fewer bits than the input. This is how the Viterbi decoder retrieves the original unexpanded data.

The decoder does not output any unexpanded data until it is satisfied it has combed errors out of the input stream. It will save up data until this is the case. The decoder looks back over the recent history of state changes to see if those changes make sense. If the history looks suspicious, the Viterbi decoder considers the fact that one or more of the received data bits may have been wrong. One by one, the decoder looks into recent history, examines each recent input bit that was received, and considers what would happen if that bit were in error. The decoder determines what the hypothetical history would have looked like in this case. If the hypothetical history looks much better, the decoder gives weight to the fact that the input bit may have been in error. Once the decoder finds the most satisfactory hypothetical history, it outputs that hypothetical data as the real, corrected, unexpanded data.

As we mentioned before, if the Viterbi decoder cannot correct all the errors, it at least collects them all in one place. Because it stores up the data history before deciding and making an output, it will output all the recent errors in a burst as it fails in its task. Even in failure, this is a key to success. The deinterleaver spreads out the burst of errors so the RS codes can correct them. Picture the Viterbi decoder giving up, but sweeping its mistakes into one corner and pointing them out to the RS decoder so they can be cleaned up there instead.

As a reminder, this version of Viterbi coding pertains to the case where the transmitting Viterbi encoder can expand the data before transmission. The encoder thus organizes the unexpanded data into expanded data that has some recognizable patterns in it. These recognizable data patterns are what the decoder looks at to determine if the received, expanded data is suspicious. The metric the decoder uses to examine data is the Hamming distance (the total number of bit differences) between the actual received data and the hypothetical received data.

The Viterbi algorithm comes built in to many standard communication systems. The single most important thing to keep in mind when using the Viterbi algorithm is that it comes in varying degrees of strength based on certain parameters. If the parameters can be varied, the stronger codes will tend to expand the data more.

The following web sites describe Viterbi coding further:

- http://pw1.netcom.com/~chip.f/viterbi/tutorial.html (follow the links)
- http://ece-classweb.ucsd.edu/Archive/winter02/ece260b/Labs/Project_2/easy_viterbi.htm
- www.cim.mcgill.ca/~latorres/Viterbi/va_main.html

Bandwidth Limited

Sometimes there is no extra bandwidth to work with, and we cannot use Viterbi coding to expand the data at all. If we want to make some of the transmitted data redundant, then we must throw away some bandwidth. A bit less of the original data is transmitted, but the bandwidth is filled up with some redundant data. The overall bandwidth remains unchanged.

The operator of the communication link must decide how much data bandwidth to give up. Trellis coding and Viterbi decoding can be used. These codes expand the data by a definitive amount based on the coding strength. The operator of the communication link can select a code and decrease the input data by just the right amount so the coding expansion fills up the channel bandwidth.

Remember, the goal of this type of encoding is to organize the input data with recognizable patterns so the decoder can determine if the channel noise has altered it. The technique generally used to accomplish this is to sacrifice some of the symbol positions in the symbol constellation.

Suppose, for example, that the symbol constellation looks like Figure 9-10 showing 64 QAM, which we've seen before. In the case of QAM that is not encoded, all transitions are possible. The signal can move from any X mark to any other X mark to signal the transmission of 6 more bits ($2^6 = 64$). But it is possible to restrict the possible transitions in a recognizable way. If, for instance, it was only possible to jump from one X mark to just 32 other X marks, then only 5 bits would be transmitted by the transition ($2^5 = 32$). The data rate would be cut in half, but the signal would have to follow a distinct set of rules that would be known to the decoder. The decoder would then be in a better position to detect errors by the means, outlined earlier.

One other technique for restricting the transitions the symbols can make is to literally provide extra symbol positions. Consider, for the moment, a 16 QAM system with the symbol constellation shown in Figure 9-13.

It's possible to double the number of symbol positions to make a 32 QAM system and to double them again to make a 64 QAM system. The symbols in the 32 QAM system can be arranged in any geometric arrangement but are best packed into an approximation of a circle (see Figure 9-14).

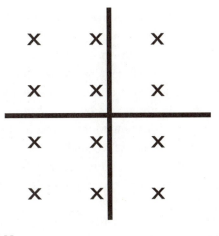

FIGURE 9-13   16 QAM

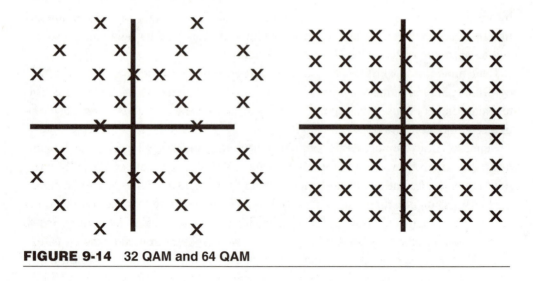

FIGURE 9-14   32 QAM and 64 QAM

If the 16 QAM system is still going to transmit the equivalent of 4 bits per symbol, the encoder can pack in redundant data by restricting the 32 or 64 different symbol locations that will be permitted transitions. This effectively puts an identifiable pattern into the data without expanding the bandwidth. The Viterbi decoder can still be used, but it uses the distance between symbols as a metric as it looks for suspicious data transitions.

Turbo Coding

Some advances have been made since Viterbi brought out his codes in the late '60s. Viterbi decoder complexity tends to grow exponentially for stronger coding gains. Classical turbo codes have been out for a while, with much better results, but the classical turbo codes have some limits, reaching a limit short of the Shannon capacity limit. In addition, the classical turbo codes are complex to compute and use expensive hardware. Turbo product coding (an improvement on classic turbo codes), is more promising. The technique allows a determined communications link designer to get arbitrarily close to Shannon's capacity limit, sending as much data through a channel as the S/N ratio will allow. A good deal of computation is required, but the computations tend to be iterative and lend themselves to an implementation in silicon. Performance is largely bounded by memory limits.

Turbo product codes replace the entire RS, Viterbi concatenated chain. The performance delivered can bring the BER curve within an arbitrarily small number of dB of the Shannon limit. The coding is similar to the block coding of RS, with data and checksum bytes, but it has several differences.

First and foremost, whereas RS has a row of data and checksum bytes, the turbo product codes have a three-dimensional structure. Checksums are computed for all three dimensions for the data: x, y, and z. In this way, the original data is given error-correcting checksums in multiple directions. The decoder has a relatively simple compute engine. The decoder works on one checksum at a time, doing x, y, and z vector checksums in separate calculations. Every time the decoder compute engine makes corrections on a vector, it changes the results in the other two dimensions. Once the decoder has been used on all the vectors in all three dimensions, the entire process can start over. The decoder can process the data as many times as needed to make the data as perfect as possible. The more times the decoder is used, the better the results. If the data is known to have many noise errors, the decoder can be used several times. If the data is known to be fairly clean, the decoder can be used one or two times. Sufficient information is built into the originally transmitted data so the decoder knows when to stop iterating through the received data.

The following web site and PDF files have further descriptions of turbo codes:

- www-ext.crc.ca/fec/Compare_Ref2.pdf
- www.ee.vt.edu/~yufei/turbo.html

INTERLEAVER

Interleaving is a way of spreading out errors. Often, an error-correcting scheme will break down if the errors occur in a regular pattern. Viterbi codes, for instance, will gather errors into concentrated bursts. An interleaver takes adjacent data and moves them apart, much like a deck of cards is shuffled. The data is not expanded, just rearranged. The encoder can interleave the data before transmission, and the decoder can deinterleave the data on reception. Interleaving can be done in many different ways, each of which conveys specific advantages and disadvantages.

Here's the bottom line on interleaving. In general, if interleaving is used within a standard communications link protocol, all the options are already specified. In this case, no choices will affect the performance of the communications link. More information can be found at these sources

- www.es.lth.se/home/jht/interleaverdownload.html
- www.comblock.com/download/com1016.pdf
- www.cs.ucl.ac.uk/staff/jon/talks/rtpi/sld001.htm

Here interleaving is used with compression, not coding. See slides 3, 4, and 5.

Shared Access

Communication links are often used by multiple communications entities: sources and destinations. Sometimes the entities are in separate computers; sometimes they are in separate processes in the same computer. If a link has multiple sources and destinations, they have to contend for the use of the communications link. Often, the physical layer will not allow them all to use the communications link at the same time. The designer of the communications link must devise a strategy that enables each communications entity the maximum amount of access to the corresponding communications entity on the other end of the link. Given that a limited amount of bandwidth exists in the communications link, the designers have to watch out for many different requirements.

BANDWIDTH

Every communication session between entities has different bandwidth requirements. The requirement for bandwidth may change over time. Some sessions will require a very steady bandwidth, and some sessions will suddenly require a large percentage of the available bandwidth. These sessions will present various types of demands on the bandwidth.

Raw Bandwidth

The different communications sessions may all have different requirements for bandwidth.

Changes in Bandwidth

Some communication sessions have bandwidth requirements that change unpredictably. On a shared environment, it often takes time to negotiate for more (or less) bandwidth. It takes time to conclude such negotiations, which time must be taken into account by the designer.

Further, the different communication sessions may all have changing requirements for bandwidth. This presents a classical problem of how to pack all the competing requirements into the available bandwidth. Even if enough bandwidth exists to satisfy the arithmetic total of all the required bandwidths, it still may not be possible to pack them together inside the channel. This problem is reduced to a classic mathematical packing problem. This problem is akin to trying to pack different-sized blocks inside a box. There's almost no way to do it without wasting space. Even if the blocks could all fit at once, there may not be enough time to determine the proper solution. The net result is that full bandwidth is rarely achieved in these circumstances. For more info, access the following sites:

- http://eng.murdoch.edu.au/EngModules/m108demo/Section01/Section0102c.html
- www.nist.gov/dads/HTML/binpacking.html
- http://mathworld.wolfram.com/Bin-PackingProblem.html

It should also be noted that some receivers have receiving buffers that must not overflow or underflow. This is true of MPEG transmission, so be sure to read the following section. It's possible for the channel bandwidth to vary because of errors. This presents much the same problem as the varying requirements. Sometimes errors must just be accepted.

Guaranteed Bandwidth

Some communication links require a guaranteed bandwidth. MPEG video data streams coming back from the robot would, in general, require constant bandwidth. Such bandwidth would have to be reserved in advance, or at least not be subject to repeated renegotiation.

Reverse Channel Bandwidth

Bandwidth is often thought of as a one-way parameter. The truth is, if the channel is bidirectional, then the bandwidth must be sufficient in both directions. This can greatly complicate systems where bandwidth is arranged at the spur of the moment.

DELAYS

Several types of delays can disrupt a communications link. All communications links have delay. Even at the speed of light, data can take microseconds to cross a county. Most electrical signals move far slower than that. Electronic boards for communication have significant processing time, which will delay data. If real-time control loops depend on a communication path, then these delays must be calculated into the design of the system.

In some systems, bandwidth is relinquished when it is not needed. Further, when bandwidth is needed, it must be requested and granted. The delay in regaining the rights to the communication channel must be added to the communication delay to determine the worst-case delay.

PRIVACY

We will discuss security and privacy shortly. The only reason to mention privacy here is that shared communication channels carry an extra risk of eavesdropping. This is especially true if all users have the option of seeing all the traffic. TCP/IP systems often have this limitation.

SHARED ACCESS ENVIRONMENT

A system in which multiple entities share the communications link can be designed in many ways. Sometimes the very nature of the communications environment dictates the methods used. Here are a couple of considerations a robot designer should take into account when picking a communications system that will support multiple entities that share access to the channel.

Closed System

If access to the communications system is restricted, then the designer can generally count on uniformity of response. The entire system should behave according to the architecture and protocols envisioned. If the communication link must be shared with unknown communication entities, then all sorts of problems can arise.

Load Limits

The total amount of communication traffic that a link will bear is often determined by both the protocol and the users' actions on the link. It is not unusual for a communication link to top out at a fraction of the raw bit speed of the link.

Cooperative Users

If the communication entities cooperate, then the usable bandwidth of the communication link can be increased. If the communication entities can be synchronized, then they can time-share a communication link fairly efficiently.

TYPES OF SHARED ACCESS

As we mentioned before, cooperation between communication entities that share a communication link is beneficial. Here are a couple of specific types of shared access arrangements that are quite general. These same types of shared access arrangements are used in many different communication standards. If a communication system is functionally identical to these systems, then the math pencils out the same way. The limits on effective bandwidth are very real.

TIME DIVISION SYSTEMS

Shared access to communications link can be accomplished by dividing the like by time division, frequency division or code division.

Aloha System

The Aloha communications system was designed so a sender could simply transmit a packet on the channel whenever it wanted to. If another sender was sending a packet at the same time, they would collide and both packets would be lost. As more and more senders compete for the channel, the system rapidly loads down. The way the math pencils out, only 18 percent of the channel's raw bandwidth is truly available once the system loads down. Normal 10BT LAN systems work like this; collisions ruin the data packets. As a 10BT LAN starts to load down with more and more users, the overall effective bandwidth of the 10BT systems is not the raw bit speed of 10 Mbps but is closer to 1.8 Mbps. On a 10BT LAN, this limit can only be improved if the users cooperate.

Slotted Aloha

The Aloha system can be improved if the senders are synchronized. Each sender knows when the timeslots occur and can only start to transmit at the beginning of a timeslot. Collisions still occur, but this sort of cooperation between senders increases the effective throughput of the channel to about 35 percent of its raw bit speed.

Reserved Aloha

If the senders politely reserve timeslots in advance, the effective throughput of the channel increases yet again. Although some bandwidth is wasted making the reservations, collisions are largely eliminated and the efficiency can be high. Only the reservation timeslots are wasted. Reservations can be granted in multiple ways, including round-robin, priority systems, and random selection. It is up to the robot designer to determine what sort of "request-grant" system to adopt.

FREQUENCY DIVISION SYSTEMS

It is certainly possible to put different communication entities on different frequencies within the allowable communication channel frequencies. Several issues arise, such as frequency allocation and frequency separation.

Frequency Allocation

All of the same reservation issues of reserving bandwidth are present in frequency division systems. If a frequency goes unused, then the bandwidth is wasted. If reservations are required, then overhead exists for making the reservations.

Frequency Separation

Communication channels on adjacent frequencies must not interfere with one another. Filters are used to remove adjacent frequencies from a communication band.

Since perfect filters are impossible to make, we must leave extra bandwidth between frequency bands. It is impossible to pack different frequency bands too close together. Both the transmitter and receiver run into trouble if they are too close.

A few other problems can crop up when frequency bands are packed close together.

- **Distortion**   The transmitter may have trouble with intermodulation distortion. Consider the case where two frequencies, f1 and f2, are amplified and up-converted together. The result is unwanted distortion signals at frequencies (f2 − f1) and (f1 + f2). Here is a PDF file and a few URLs speaking about such distortion:

    - www.sinctech.com/pdfs/Intermod.pdf
    - www.audiovideo101.com/dictionary/im-distortion.asp
    - www.atis.org/tg2k/_intermodulation.html

- **ISI**   If frequencies are too close together, the electronics handling each frequency may have trouble filtering out the adjacent signals. Although frequency division systems are viable and work fine, time division and code division systems have stolen the thunder of this technology.

CODE DIVISION SYSTEMS

Code division systems use a form of encryption where each user's data is invisible to the other users.

Code Division Multiple Access (CDMA)

CDMA systems, also known as *spread spectrum* (SS) systems, generally use a wide frequency bandwidth. The data for each user is spread across the entire frequency band using a spreading code. Every user's communication is broadcast in the band at the same time, but they do not interfere with one another. Each user gets a unique spreading code that is used to separate users. Because of the nature of the codes, little or no interference exists between users. Furthermore, no synchronization is required between the users.

In 1940, Hollywood actress turned inventor Hedy Lamarr copatented a frequency-hopping device for military use. It's kind of nice to have her picture in here among all the men in powdered white wigs (see Figure 9-15). One interesting quote is attributed to her: "Any girl can be glamorous . . . all she has to do is stand still and look stupid." More information on her work can be found at www.inventions.org/culture/female/lamarr.html and at www.edu-cyberpg.com/Teachers/womenmonth.html#ahedy.

Here's how the most popular SS systems work. Each user is assigned a coded spreading waveform:

$$U(code_i)$$

FIGURE 9-15 Hedy Lamarr, pioneer in spread spectrum communications and actress

Where $code_i$ is the user's unique code that selects the characteristics of the waveform $U(code_i)$. These waveforms are typically a series of pulses that have the following characteristics. Whereas $\times$ represents a bit by multiplication (correlation):

$$U(code_i) \times U(code_k) = Z \ll 1$$

unless k = i and the two waveforms are synchronized, in which case

$$U(code_i) \times U(code_i) = 1$$

In addition, $U(code_i) \times B$ is very small for uncorrelated signals (like radio transmissions) that may already exist in the channel. This means that SS signals can coexist (overlay) in the channel with existing communication users.

In some SS protocols, the data is first modulated by the spreading waveform prior to transmission. Consider the case where the channel is filled with the waveforms of two users. We can extract a single user in the following way:

$$Channel = D_i \times U(code_i) + D_k \times U(code_k)$$

where i and k are different and D_i is the data from user i.

$$Channel \times U(code_i) = U(code_i) \times (D_i \times U(code_i) + D_k \times U(code_k))$$

$$Channel \times U(code_i) = D_i \times U(code_i) \times U(code_i) + D_k \times U(code_k) \times U(code_i)$$

$$Channel \times U(code_i) = D_i + Z \cong D_i$$

Similarly, the waveform for user k can be cleanly extracted as well.

On the plus side, SS communications can coexist with existing, uncorrelated communication signals in the channel. This basically allows the channel spectrum to be reused.

On the minus side, the different codes are not completely orthogonal. The previous small signal Z is multiplied by the number of other users and can interfere with reception. This can limit the number of users.

Here are a few PDFs discussing shared access communication links:

- http://courses.cs.vt.edu/~cs5516/spring02/Phy_mac_6.pdf
- www.cse.sc.edu/~srihari/csce516/lecnotes/shared.6.pdf
- http://web.mit.edu/course/16/16 .682/www/lec18.pdf

Compression

Often, the bandwidth available for digital communication is limited. This may occur for several different reasons:

- **Regulated spectrum**   The government may regulate access to the spectrum and make everyone share it.
- **Cost**   It is often too expensive to purchase rights to the needed spectrum.
- **Energy**   As we discussed before, sending bits across a wireless channel literally requires a sufficiently high Eb/No. In satellite transmission, this fact literally comes home as satellite batteries and solar panels struggle to provide energy to each and every bit. Robots in remote locations are often up against this very problem. Don't forget one thing though. It takes energy to compress the data in the first place. The compression process may have to be very energy efficient and the entire process will have to be analyzed.

Whatever the reason, there is little point in sending useless data across the channel. Most digital communications can be compressed to a smaller amount of data. Shannon toyed with this at some length. To test this assertion, pick a few different types of files on a computer and try to compress them with WinZip.exe, a trademarked program from WinZip™ Computing, Inc. It's presently available for trial use at www.winzip.com/ddchomea.htm.

The following compression rations can often be achieved:

- **Standard text files**   A factor of 6 to 10.
- **Program files**   A factor of 2.
- **Video or graphic files**   A factor of 1.1.

Try compressing these types of files with the WinZip™ utility to see what can be achieved. If the robot's digital communications must be compressed prior to transmission, several options are available. WinZip™ may not be usable on the robot's computer. It's likely that the robot's operating system software library (or freeware) may already have compression utilities that can be used. Two basic types of compression are commonly used by these programs. These techniques are used in standard compression programs and can be rewritten to suit the needs of the robot.

Fourier Transforms

Graphics and video transmissions are routinely compressed using Fourier transforms. In MPEG video compression, the pictures are converted to a series of coefficients that are compressed again using run-length compression (described later). Compression ratios of 50:1 can be achieved. It is not simple to write (from scratch) a program to perform this type of compression. Here are some web sites and a PDF discussing image compression:

- http://it.wce.wwu.edu/jongejan/461/Video.html
- http://arachnid.pepperdine.edu/grosenkrans/compression.htm
- http://poseidon.csd.auth.gr/LAB_PUBLICATIONS/Books/dip_material/chapter_4/main.htm

Run-Length Compression

One of the oldest, and most intuitive, techniques of compression is simple run-length compression. Instead of sending, for example, a series of 2,415 zero byes, we can simply send a block of data that is about 4 bytes long, explaining it represents 2,415 zero bytes. The protocol is simple and can be written from scratch if need be. The following URLs explain a few different types of run-length encoding:

- www.rasip.fer.hr/research/compress/algorithms/fund/rl/
- http://datacompression.info/RLE.shtml

Huffman Compression

Huffman compression can be used if the data can be broken up into symbols (like text bytes). Then all the symbols are reassigned a different code before transmission. The most often transmitted symbols are assigned short codes. Symbols that are rarely transmitted are assigned longer codes. Run-length coding is also applied. The following web pages are interesting lectures on data compression in general:

- www.eee.bham.ac.uk/WoolleySI/All7/body0.htm
- www.cc.gatech.edu/~kingd/comp_links.html
- www.eee.bham.ac.uk/WoolleySI/All7/links.htm

Encryption and Security

Instances occur when data must be encrypted before it is transmitted. The physical links of most communication channels move through areas that are in public. Certainly, all *radio frequency* (RF) communications that move through free space can be intercepted, and phone communications move through common wiring and facilities. We've all heard of the infamous hackers who eavesdrop and create other problems on the Internet. We'll use the term hacker to refer to unauthorized parties who may be up to no good. Hackers have a variety of motives and any hacker would love to gain control of a robot.

Just like Internet communications, RF transmissions and phone traffic are also subject to interference by hackers. The transmitted data can be read or altered en route. Here's a list of the things that can go wrong when a hacker is involved:

- ***Denial of service* (DoS)**   If a hacker jams the communication link, commands to the robot may not get through.
- **Eavesdropping**   Hackers may read the robot's data and get vital information they may be able to use.
- **Spoofing**   Hackers may pose as the source or destination communication entities. The following problems could then arise:
  - **False commands**   The robot may receive false commands and execute operations that could damage the mission.
  - **False data**   Data could be falsified or altered. The integrity of scientific studies and data-gathering missions could be compromised.
  - **Broken communications**   If the hacker succeeds in spoofing the other communication entity, the entire communication chain may be interrupted going forward.

The following web sites are hacker clubs that you can visit at your own peril. Your computer may be threatened and strong language may be involved. There may, however, be things to learn. Hackers know more about security than most people. At the very least, they may scare your socks off:

- www.phrack.org
- www.morehouse.org/hin/

Security and hacker problems can be solved in multiple ways. Several standards have been set up to encrypt and harden communication links to prevent hackers from intervening. Some of these methods are more effective than others. Most casual hackers are harmless, but enough determined hackers are out there to crack almost any code. As a rule of thumb, ask what would happen if a hacker had full access to the communications link. What's the worst that could go wrong? If your data is boring or nonvital, then don't bother. However, minimal security is often beneficial and does not cost much. But remember, robots are an attractive technology target. The prospect of hacking into a robot will wake up even the sleepiest hacker.

In a nutshell, what sorts of fixes are available? Eavesdropping becomes almost a useless exercise if the data is completely encrypted. Further, spoofing becomes impossible if the message has sufficient authentication to verify the sender. Methods to provide encryption and authentication are outlined in the following URLs. Many computer software libraries contain subroutines to support secure communications. If the data stream moves too fast for software encryption, hardware chips are available that can encrypt the information faster. Popular encryption standards are listed at www.cs.auckland .ac.nz/~pgut001/links/standards.html and include the *Data Encryption Standard* (DES), RSA, and *Pretty Good Privacy* (PGP)™.

DATA ENCRYPTION STANDARD (DES)

DES has won the backing of the government and is present in many commercial transactions today. The calculation methods are fairly straightforward and chipsets are available for high-speed implementations. Further information on DES can be found at the following URLs:

- http://axion.physics.ubc.ca/crypt.html#aDES
- www.tropsoft.com/strongenc/des.htm
- www.tropsoft.com/strongenc/des3.htm

RSA

RSA security is based on the fact that it's very difficult to factor large numbers. If a hacker could factor a huge number in less than a few years, the hacker could break into the communications link. So far, it's proven too difficult (see www.rsasecurity.com/).

PRETTY GOOD PRIVACY (PGP)™

PGP™ security is also based on difficult mathematical calculations and is offered in several versions, as detailed at the following sites:

- www.pgp.com/
- www.neiu.edu/~ncaftori/PGP.htm
- www.scramdisk.clara.net/pgpfaq.html

Dos attacks happen when a determined hacker sends packets to the robot that it cannot handle. Some DoS attacks involve sending packets with an illegal data structure; other DoS attacks involve sending too many packets so the channel gets clogged up. Be sure your software can handle packets with illegal structures, and consider testing it with simulated DoS attack data. Some web sites speak to this issue, such as www. geocities.com/solarsistem/gif/docs/dos.htm and www-arc.com/sara/coe/distributed_ denial_of_service.html

The following web sites contain methods that can be used to secure the transmission into and out of the robot:

- www.postech.ac.kr/cse/hpc/research/webcache/book/security/total.html
- www.cs.auckland.ac.nz/~pgut001/links/standards.html
- www.cs.auckland.ac.nz/~pgut001/tutorial/
- www.11a.nu/security.htm

Popular Communication Channels

In designing the robot, it makes sense to stick to tried and true communication protocols. Several protocols, for both wireless and wired communication channels, are available and popular. Usually, this means that the hardware and software can be purchased off-the-shelf. Robots not only need to be designed quickly, but they need to be reliable. So make sure you check the pedigree of any commercial product. It's easy to say that a product conforms to a standard, but most of these standards are so complex that the newer offerings are not as reliable as the older, more established products.

WIRELESS SYSTEMS

Several wireless data systems have been deployed for some years. We'll discuss some of the features and the performance of each one.

Wireless Fidelity (Wi-Fi)

Wireless fidelity (Wi-Fi), or 802.11b, is a wireless version of Ethernet LANs. It uses RF communication in the 2.4 GHz band and the protocol is documented in 802.11b. Many people have cards in their laptop PCs that can tap into Wi-Fi portals in stores, businesses, and public places. This technology makes Internet access available to portable computer users. It would make a fine communications link for a robot as long as security and other issues are handled properly.

The data bandwidth is similar to that of 10BT wired LANs that we discussed earlier (about 1.5 Mbps or so). The range of transmission is up to a couple hundred feet, but data speeds can drop after 50 feet. The protocol uses spread spectrum communications, as discussed earlier. As such, it is adept at overlaying an existing communication spectrum and coexisting with the communications traffic in it. Faster versions of the protocol are just coming out now. For more information, go to www.wi-fi.org and http://alpha.fdu.edu/~kanoksri/IEEE80211b.html.

General Packet Radio Service (GPRS) Data

The worldwide *General Packet Radio Service* (GPRS) system supports data transmission. Although the frequency in the United States is different than overseas, the data-carrying capability is similar. The GPRS data system uses RF communications in the 800 to 960 MHz bands and the data bandwidth is at most 170 Kbps , but in practice, it's best to limit expectations to one-tenth of that. The range of transmission is similar to cell phones and is subject to similar blackout zones according to geography. The protocol uses *Global System for Mobile Communications* (GSM) and *Time Division Multiple Access* (TDMA) communications with narrowband GMSK-modulated communications and *Time Division Multiplexing* (TDM). As such, users can negotiate for more time slots and higher data bandwidth without a loss of accuracy. More info can be found at www.gsmworld.com/technology/gprs/intro.shtml#a1a and at www.ieng.com/warp/public/cc/so/neso/gprs/gprs_wp.htm.

Bluetooth

Bluetooth is an RF channel that is somewhat new and just coming into its own. It, too, uses RF communication in the 2.4 GHz band and GFSK (frequency shift keying). It uses spread spectrum communications (via frequency hops) among 79 different 1 MHz-wide bands. It is meant for short, 10-meter-range communication. The data bandwidth can be as fast as 723 Kbps, but practical limitations restrict the bandwidth to about half that. This makes it poor for video, but good for Internet communications. It is adept at overlaying an existing communication spectrum and coexisting with the communications traffic in it. Check out www.bluetooth.com and www.csr.com/enews/sw007.html for further information.

Infrared Data Association (IRDA)

The *Infrared Data Association* (IRDA) standard link is an infrared channel that has been around for some time. It uses infrared light for short-range communications of about 1 to 2 meters. It uses baseband frequencies modulated up to 1.5 MHz to transmit data at up to 4 Mbps. It is commonly used for short communication sessions between computer peripherals. The following web site and PDF files provide further details on IRDA:

- www.irda.org/
- www.irda.org/use/pubs/Overview.PDF
- www.irda.org/design/infrared_data_communications_with_irda.pdf

WIRED SYSTEMS

A few wired communication systems are widespread at this time. They can all support higher-level protocols, so we'll start with just the physical layers.

Phone Network

The common carrier phone systems can be used to transmit data, which can be done in a couple of different ways:

- **Dialed services** Several companies (like AOL) have phone numbers that a robotic computer could dial up to access the Internet. Modems are required to support connections to the phone line. These modems have a top end of 33 to 56 Kbps. The top speed will usually depend on the quality of the phone connection. Service can be denied if the line is busy. (See www.driverzone.com/guides/modem/intro/modemguide_p3.html and www.v90.com/v90magic.htm.)
- *Digital subscriber line* **(DSL)** Given that one phone wire already comes into the house, the phone companies use frequency division to put DSL signals on the wire. Voice traffic only uses frequencies below 10 kHz (at most). DSL signals typically use the same wire to carry QAM signals at bit rates around 1 Mbps. The service is continuous and largely based on the Internet. Service may be interrupted if the robot does not exercise the communications link now and then. The phone company takes away the robot's IP address if the robot goes idle for too long. Just make sure the robot is active now and then so it keeps its IP (DHCP) reservation intact. More info can be found at www.howstuffworks.com/dsl.htm and at www.dslforum.org/.

Cable Networks

Many homes also have cable TV coming into the house. Although the cable system was originally designed as a one-way system, many of the cable systems now have reverse channels capable of taking information from the homes back to the cable company.

The standard that most cable TV companies use is the *Data Over Cable Structured Interface Standard* (DOCSIS), which provides time division access to home subscribers. The cable system is a closed system so the physical layer can be proprietary. The modulation methods used are *Quadrature Phase Shift Keying* (QPSK), 16 QAM (upstream), and 64 QAM (downstream), backed up with RS coding, as discussed earlier.

It should theoretically be possible to buy any DOCSIS modem and use it on the cable system, but this may not always be the case. DOCSIS provides data rates downstream up to 56 Mbps and upstream rates up to 3 Mbps. However, all the data bandwidth must be shared among all the users (and the sideband control information). More information is provided at www.cable-modems.org/tutorial/01.htm (follow the page links) and at www.iec.org/online/tutorials/cable_mod/ (follow the page links).

Local Area Networks (LANs)

One of the most popular wired communication systems for computers is the LAN and the Internet. A LAN is a method of connecting computers together in a building or small campus. The Internet is the network connecting computers together worldwide. We'll take a look at the physical layer first and then discuss some of the basics needed to plan a LAN connection for the robot.

Physical Layer   The most popular method of connecting to a LAN is the Ethernet. Most computers have NIC cards or connections that can accept Ethernet connectors at 10BT/100BT data rates. 1000BT (and variants thereof) provides 10 times the bandwidth, but we'll ignore it for now. The top end of the commonly available data bandwidth is 100 Mbps with 100BT. But as we discussed earlier, the Ethernet LAN is a slotted Aloha system without reservations. As such, uncoordinated traffic tends to top out at 18 percent of the raw capacity of the network. The transmission method is baseband NRZ data, as discussed earlier, but the tough part is not the physical layer.

The toughest part of using a LAN is the configuration. It takes an expert to tame a LAN system, so plan on consulting heavily with *information technology* (IT) personnel before deciding on a LAN for the robot. They'll ask tons of questions before setting up the LAN so the robot can use it. They'll provide the IP address the robot needs to function as well as the connectivity to the other computers the robot will need to address. In addition, they can set up the robot with other services it may require, like email and Internet access.

We must cover a couple of basics before turning the robot loose on the LAN. Books that explain how to effectively use LANs are often a thousand pages long. The following are the basic facts most often used in designing a LAN communication link for a robot. These facts do not sufficiently explain all the details of how to finish the engineering on the LAN link, but they outline the capabilities of a LAN so the robot's communication link can be planned.

TCP (Error-Free) Communication

If error-free data transfers are required, the data session on the LAN must be set up with a socket connection. Every computer attached to the LAN (and Internet) has a specific address. Most of the communications will be between the robot and a specific computer (point to point). The connection must first be established before transfers can take place. Most computer software operating systems have stacks (modeled after the OSI stack) that assist in the formation of all the connections that are needed.

Although the software is not difficult to write, it makes sense to get an experienced software engineer to write LAN communication software. If an inexperienced person writes the software, it will have errors that will cause significant problems later. The programmer will bring up the software stack, obtain an IP address, establish the connection (termed a socket), transfer all the data, and close the socket at the end of the session. The following web sites outline the connection process:

- www.cs.rpi.edu/courses/sysprog/sockets/sock.html
- www.ecst.csuchico.edu/~beej/guide/net/html/
- http://java.sun.com/docs/books/tutorial/networking/sockets/
- www.exegesis.uklinux.net/gandalf/winsock/ (for Windows)
- www.cs.berkeley.edu/~kfall/EE122/lec23/sld001.htm

User Datagram Protocol (UDP) Connectionless Communications

Sometimes error-free communication is not required. In fact, sometimes it is impossible. Consider video communications. Because video receivers require a constant stream of data, no time is available at all to go backwards and retransmit the information in error. Instead, the video screen simply freezes for a while until new data comes along. The sending computer need not forge a connection in advance. It can simply determine the IP address of the receiving computer and start transmitting. The receiving computer need not send signals back at all.

Broadcasting

In fact, if the IP addresses are set up properly, multiple computers can receive the transmitted data (also called UDP datagrams) at the same time. This technique is called *broadcasting* and is a useful way to ask around for information. The sending computer can ask via UDP broadcasting, for instance, if any other computer has specific information. Computers wanting to reply can establish a TCP connection to reply privately.

The only restriction on broadcasting is that it tends to stop at the boundaries of a LAN. Broadcast data cannot be allowed out on the Internet because it would flood the system. UDP Broadcasts must be kept inside a LAN non-UDP. Broadcasts to multiple locations on the Internet are often set up inside a server using multiple point-to-point connections and simultaneous transmissions of the same data.

Be careful of a couple things when using UDP communications. First of all, the data will not be error free. Second of all, the packets may not even arrive in the right order. TCP takes care of such things. In UDP communications, if such things are important, they have to be taken care of in the application software written for the robot. A web site explaining LAN technology is at http://punch.engr.wisc.edu/~orchard/net-tutorial/.

Okay, you've been so patient learning communication techniques that you deserve a reward or two for getting this far. Don't tell anyone else this; they have to read this far to get it! After receiving a complaint that he was ending a sentence with a preposition, Churchill said: "This is the sort of pedantry up with which I will not put." Check out the following web site: www.winstonchurchill.org/quotes.htm#put.

The Voice of the Robot!

The following is the voice of the future: a text-to-speech engine that illustrates just how far the technology has come in the last few years. I suggest going to http://eserver.org/history/gettysburg-address.txt. Copy just the first two lines (more than 30 words) from Lincoln's speech, paste them into the text box at www.research.att.com/~ttsweb/cgi-bin/ttsdemo, and submit it for processing. The results are great fun. Pick the voice you like best. Personally, I find the results amazing.

MOTORS AND ACTUATORS

Motors are simply devices that take in power and generate movement. Most motors convert the power to a magnetic field using coils. A few motors do not use coils, and we'll discuss them later.

The power fed in to the motor coils can come from the AC power mains, DC power supplies, or from controllers that control the coils for specific purposes. Motors are divided into classes based on the type of power they use.

AC Motors

Most motors in use today are AC motors designed for medium to heavy-duty work. They are present in most motorized appliances that use AC power. They are inexpensive because they do not require complicated construction and because they are built in large quantities. Motors differ in their construction, speed control, cooling methods, control systems, size, and weight.

- **Construction** AC motors have the coils built in to the outside casing (the stator) and magnets that spin in the middle (on the rotor).

275

- **Speed** The number of windings and the frequency of the power fed to the coils fix the speed of the motor. The speed of AC motors is basically constant. As such, they may not be the best for robots. Let's consider just 60 Hz of power for these examples. If just three windings form a single rotating field (one pole), the motor spins at 60 Hz or 3,600 *revolutions per minute* (RPM). As three more winding coils are added, the number of poles goes to 2 and the RPMs go down to 1800. The following equation is used to determine the RPM, where p is the number of three winding coils (poles), f is the frequency of the power, and s is the speed of the motor in 4 RPM:

$$s = 60 \times f/p$$

- **Cooling** The windings are on the outside case, where they can be cooled easier. Furthermore, with no brushes, the casing can be wide open to admit air for cooling.
- **Controls** AC motors are not easy to control, in either speed or position. It is possible to build an electronic controller to trim the speed and power consumption of an AC motor, but it is best used in situations where only gross mechanical power is needed, especially for constant speed applications.
- **Portability** Given that a portable robot probably is running off batteries, AC motors may not be the right choice. Along with the difficulties of controlling the speed and position of an AC motor, it's fair to conclude they may not be a good choice in a robot.

DC Motors

DC motors come in many different styles. AC motors only have fewer styles because their architecture attempts to take advantage of the existing movement (waveform) of the AC power.

Like most motors, DC motors generate movement by creating magnetic fields within the motor that attract one another. By and large, DC motors have permanent magnets in the stator and the rotor has the coils (the reverse of AC motors). But since DC power has no movement (waveform) of its own, the motor electronics must create a change in the DC waveform as the motor rotates. This can happen in several ways.

DC MOTORS WITH BRUSHES

- **Construction**   The rotor would stop spinning if the DC field in the rotor coils never changed. By altering the polarity of the DC voltage on the coil as it rotates, we can continually make its field attract the next magnet in the stator. As the rotor rotates, a set of position-dependent switches in the rotor switch the field on the rotor coils. The switches are implemented with a stationary, partitioned slip ring on the rotor bearing (for incoming power) and brushes that drag around the ring to power the coils. After the rotor rotates enough, the brushes move to the other part of the slip ring and reverse the polarity on the coils. It's a little like keeping a carrot in front of a horse. This structure, however, has some clear disadvantages:
  - **Electrical noise**   The brushes create sparks, which emit a great deal of electrical radiation. Further, since the voltages change abruptly, the power supply noise can be severe.
  - **Fire hazard**   Sparks can touch off explosions.
  - **Reliability**   Brushes can wear out and get clogged with dirt. After a while, motors may need replacement brushes.
- **Speed**   DC motors are controlled by varying the voltage on the DC power supply. Higher-voltage motors are generally more powerful.
- **Cooling**   Cooling is a little more of a problem with DC brushed motors since the electrical coils are inside on the rotor. Furthermore, since the speed is controlled by linearly varying the power to the coils, the dissipation in the power supply can become a problem.
- **Controls**   By controlling the voltage and current through the coils, both speed and torque can be controlled. By and large, most DC motor controllers use a chopping waveform to control the average DC voltage (as opposed to a linear regulator). By turning the DC coil voltage off and on (to full voltage) very rapidly, the average DC voltage on the coil can be controlled by means of a duty cycle. Such motor drives are more efficient.
- **Portability**   DC motors tend to take up more room than AC motors of similar power because of the brushes and coils on the rotor. Further, since the coils are on the rotor, they have a considerable gyroscopic effect. A lot of spinning mass exists on the rotor.

BRUSHLESS DC MOTORS

- **Construction**   Brushless DC motors have much the same construction as AC motors. The rotor has permanent magnets, and the coils are on the case (stator). By altering the polarity of the DC voltage on the stator coils as the rotor rotates, we can continually make its field attract the next magnet in the rotor. As the rotor rotates, electrical controls switch the field on the stator coils. This structure has some clear advantages:

- **Electrical noise**   Much less electrical noise exists than with brushed DC motors.
- **Fire hazard**   No sparks are made.
- **Reliability**   No brushes are used that could wear out. Further, far less mass takes place on the rotor.

- **Speed**   DC motors are controlled by varying the voltage on the DC power supply. Higher-voltage motors are generally more powerful.
- **Cooling**   Cooling is easy since the coils are on the casing, but because the speed is controlled by linearly varying the power to the coils, the dissipation in the power supply can become a problem.
- **Controls**   Brushless DC motors can be controlled with a similar type of chopped waveform control that the brushed DC motors use (with accommodations for the interference from the brushes). Since no brushes are used, the controller must also sense the motor position. This makes the controller much more expensive.
- **Portability**   Brushless DC motors are fairly lightweight, but the controller can be complex. Further, make sure the motor does not have delicate sensing wires (to sense position). Try to get the kind where the controller senses the motor position automatically. It makes the controller more expensive, but the motor will be more mechanically reliable.

DC STEPPER MOTORS

- **Construction**   Stepper motors have much the same construction as AC motors and DC brushless motors. The rotor has permanent magnets, and the coils are on the case (stator). By altering the polarity of the DC voltage on the stator coils as the rotor rotates, we can continually make its field attract the next magnet in the rotor. As the rotor rotates, electrical controls switch the field on the stator coils. Some clear differences exist between steppers and DC brushless motors:
  - **Stepping speed**   Stepping motors are designed with more rotational positions and tend to step from position to position faster. They're more like a digital system and the DC brushless motors are more like an analog system.
  - **Stopping**   Steppers are designed to stop on a dime and hold their position. For this reason, they tend to have less rotational mass. DC motors can perform the same feat but must have carefully designed servo systems to sense and hold their position. Steppers hold the position that is defined by the motor geometry.
- **Speed**   Steppers are not necessarily designed for speed. If they go too fast, they may lose their position by slipping over one too many poles. They have to move

deliberately. They are also not well geared for changing loads; they can lose track of their position if the load varies in a sudden manner.

- **Cooling**   Steppers can be fairly open and easy to cool. If they remain stationary for some time, the current in the coil can be reduced. A good controller will do that automatically.
- **Controls**   Steppers have relatively complex controllers. They are generally computerized since the computer must keep track of the position and momentum of the motor. More complex controllers have more than just on-off control of the coil voltage and current.
- **Portability**   Steppers tend to be lightweight and fairly sturdy. They are not particularly good with large or varying loads, but they function reliably in most applications.

Exotic Motors

PIEZO-ELECTRIC MOTORS

Piezo-electric materials are ceramics that change shape when an electric field is applied across them. Watch alarms and phone ringers are the most common applications of such materials. They don't move much, but they can move often. They are used for small motions like creeping and fine adjustments. If the robot must have very fine, accurate positioning, piezo-electrics can provide the movements. They can move large loads, albeit slowly. More info on these motors can be found at the following web sites:

- www.piezo.com/intro.html
- http://web.umr.edu/~piezo/

ORGANICS

Some organic crystals expand and contract when a current is passed through them. No simpler motor exists. Unfortunately, these tend to be very fragile.

Here are some more web sites with information about motors:

- www.instantweb.com/o/oddparts/acsi/motortut.htm
- www.slewin.clara.net/elec/tmotor.htm
- www.cs.uiowa.edu/~jones/step/
- www.motionnet.com/cgi-bin/search.exe?a=cat&no=1708

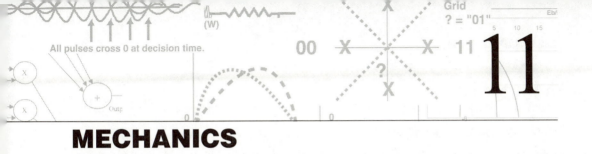

MECHANICS

The executioner's argument was, that you couldn't cut off a head unless there was a body to cut it off from . . . The King's argument was, that anything that had a head could be beheaded.

> —Speaking of the Cheshire Cat's smiling head,
> from Lewis Carroll's *Alice's Adventures in Wonderland*

So now we've spoken of energy, software, reliability, management, signals, and data. Mighty ephemeral stuff all that. Lest we forget, robots must be made of bone and gristle! Rubber, steel, plastic, fiber, and ceramics are the true stuff of robots. Just as many electrical engineers fancy themselves to be great mechanical designers, I'm still waiting for my invitation to deliver the keynote address to the American Society of Mechanical Engineers. All that said, I can pass along some tips and tricks.

Materials

Robots can be made out of just about anything. The environment and the mission of the robot often pose severe constraints on the materials that can be used. The Air Force is hoping to make robot butterflies for reconnaissance. Although it's true that most cars are made out of steel, I doubt a steel butterfly would get very far.

Many different materials are available for new robot design, and many considerations must be made when choosing the materials:

STRENGTH TO WEIGHT

Unless a mobile robot is to be used in sumo wrestling matches or very high winds, it makes sense to keep the weight down. One way to accomplish this is to minimize both the amount and density of the construction materials. We have to know the strength of the material before minimizing the amount used. Shaving material off structural members can be a risky game. It requires advanced knowledge of structural engineering and simulations. Picking a material that is not very dense is somewhat simpler. The key parameter to look at is the strength-to-weight ratio. Materials that are very strong for their weight helps keep the weight of the robot down. The selection of such materials is somewhat complicated by the fact that certain types of strength depend upon the shape and formation of the material used. For example, a well-folded cardboard structure can hold far more weight than a single piece of cardboard.

An article on the effect of new materials in designing sports equipment can be found at www.tms.org/pubs/journals/JOM/9702/Froes-9702.html.

These following web sites and PDF file outline the characteristics of various types of materials:

- www.radshape.co.uk/ (For metals, look under materials.)
- www.robotstore.com/download/Muscle_Wire_FAQ_V3.pdf (Actuator materials)
- www.mdacomposites.org/materials.htm (For composite materials)

MACHINING AND FORMATION

It doesn't do much good to have a very strong material if it cannot be formed into the shapes required for the robot.

Metals

It's easy to find a metal-forming shop in most cities. Further, metals are not difficult to work with at home; just make sure all safety precautions are observed. Very hard metals can be difficult to use because they require (expensive) tools that are even harder. Diamond studded (and other) tools are available for such work.

Plastics

Plastics can be molded, melted, and altered by machine to change their shape. Molds are expensive, in general, to build. Vacuum forming is a reasonable alternative for making thin sheets of plastic into the required shapes. Plastics can undergo machine work, even at home, but the material tends to foul the tools.

Composites

Composites can be used much like fiberglass. They can be difficult to control but offer very good strength for their weight. Metal, wood, and plastics are relatively well known materials. Composites, on the other hand, are newer and are just finding their way into consumer products like bicycles. These materials are built very much like fiberglass. Fibers, in the form of woven mats, are impregnated with a filling material that reinforces the fibers. The strength is largely in the direction of the fibers, not across the fibers. Some care must thus be taken in the design and layout of the fibers within the robot chassis. Common fibers used include glass, carbon fibers, and plastic fibers. Many different resin materials exist, such as epoxy and polyurethane. Filler materials, like short glass fibers, are available to make the resin stronger as well. The Market Development Alliance web site contains some good definitions of other materials' mechanical properties at www.mdacomposites.org.

The previously mentioned Market Development Alliance web site and the Composites Fabricators Association (www.cfa-hq.org/) are great sources of information about composites.

Wood

Wood is easy to work, but not very strong. Watch out for termites!

COST

Most of the cost of materials will be related to the machining costs. Materials, except for very hard metals and special composites, do not cost all that much.

AVAILABILITY

Metals, wood, and plastics are relatively easy to procure. Composite materials are not that hard to find either. Most of these materials can be purchased in preformed shapes like pipes, sheets, spheres, and so on. Consider starting the design using preformed parts; they can offer great strength and accuracy.

STRENGTH

Materials have several different characteristics that quantify their strength under various kinds of loads.

Tensile

This is a rough measure of how strong the material is when stretched (like a string). Glass and composite materials excel at this.

Compression

This is a measure of how well the material can hold up weight resting on it (like a post). Metals excel at this.

Flexing

This is a measure of how the material deforms with sideways pressure. In some designs, the material must not bend at all. In other designs, the material must bend (like a tree withstands a strong wind). As such, failure can come by bending too much, by breaking, and by failing to return to the original shape. All the materials mentioned can be used (by proper design) to suit the requirements of bending or nonbending applications.

Shock

This is a measure of the material's ability to survive a sudden shock. Shocks present a sudden increase in pressure on a material and radiate shock waves in ways that slow, steady pressures do not.

Abrasion

This is a measure of the ability to withstand repeated rubbing and use. Some materials will not abrade much at all. Others will not only abrade but shed harmful particles as well.

Creep

Materials subject to steady pressure will tend to give over time, or creep. Plastics, starting their life as liquids, are subject to creep. For much the same reason, metals can creep some. Just make sure that the tolerances of the robot will be maintained over time in the face of creep.

So which materials should be used in a robot? All the aforementioned factors have to be considered, but here are some guidelines based on applications:

- **Home project**   If the robot is a home project, aluminum is not a bad choice. It's cheap, easy to get, lightweight, easy to alter by machine, and relatively strong.
- **Industrial floor**   If the robot is for nonmobile industrial use, consider steel for its durability. If the volume of manufacture is high enough, consider plastics.
- **Consumer**   If the robot is for commercial release, consider plastics.
- **Space**   If money is less of a problem than weight and strength, consider the more exotic metals like titanium and composites. Many new considerations come into play for space-born robots that must face severe G forces, extreme temperature ranges, vacuum, radiation, and so on.

Some Cautions

The choice of materials can introduce other problems. A few are mentioned in the following sections.

DISSIMILAR METALS' GALVANIC CORROSION

It's never a good idea to put dissimilar metals into a robot, at least not if they come into contact with each other. Action at the atomic level can set up currents and cause corrosion. This is a particular problem in the marine environment where salts can get at the metal junction. Do not forget that fasteners must be taken into account as well. If dissimilar metals must be used, consider metal plating to decrease this effect. See the following web sites for further information:

- www.seaguard.co.nz/corrosion.html
- www.engineersedge.com/galvanic_capatability.htm
- http://corrosion.ksc.nasa.gov/html/galcorr.htm

FATIGUE

Most materials suffer damage when they are bent or otherwise deformed. Even if they return to the original shape, the damage still exists. With repeated bending, the material will eventually give way and fail. During the design of the robot, evaluate all the repeated operations. Make sure none of the materials will be stressed beyond their limits of fatigue. Consult companies that specialize in bendable materials of the type required.

CORROSION

We've already spoken briefly about corrosion in a few places, including Chapter 4. Materials can be clad in plastic or plated with other metals to decrease the rate of corrosion. If corrosion is a strong possibility, consider using materials that will not corrode. The Kennedy Space Center offers information on the causes and prevention of corrosion at the following sites:

- http://corrosion.ksc.nasa.gov/html/corr_fundamentals.htm
- http://corrosion.ksc.nasa.gov/html/publications.htm

LUBRICATION AND DIRT

Moving parts, especially bearings, sometimes require lubrication. Just remember, the basic function of oil and grease is to smear all over everything!

A buildup of grease and dirt can engender a host of problems.

- **Electrical problems** Lubricants can coat electrical contacts and insulate them from the mating contact. These sorts of failures are common.
- **Dirt** Lubricants trap dirt, causing extra friction and sluggish action. Eventually, the dirt swamps out the positive effects of the lubricant. If the robot cannot be serviced, this becomes a critical problem.

In the design of the robot, try to find sealed bearings and other moving parts that do not require lubricants. If a lubricant must be used, find an exotic one that is a bit tamer. Graphite and Teflon are possibilities, but each have their own faults.

TOLERANCES

In most mechanical designs, the parts must fit together cleanly. Moreover, the parts must continue to fit as the robot gets older. One of the most difficult tasks in building a robot is making it sound. Parts that bend and screws that come loose can make a design degrade rapidly. Such mechanical failures are probably the single worst problem plaguing robot designs.

Here's one small example of a trick that may help. Consider a three-part robot with parts A, B, and C. Also, assume all fasteners have some play that increases over time. Let's call the typical play T millimeters; the unintended movement that can occur because of inexact mechanical tolerances. Another common term for this is *slop*, although I suspect the robot would be offended. Although this is a gross oversimplification (and in one dimension), it can be used to illustrate the design of tolerances. Here are two ways a design can be built under these conditions.

- **Bad design** A bad design would attach A to B, and B to C. Part C will move with respect to part A with movements that could be the sum of the other two tolerances, or 2T. The other two pairings will move respectively within the tolerance T.
- **Good design** A good design would attach A to B, B to C, and A to C. Slop within the system will be limited to roughly T, not 2T.

In general, consider having a central, rigid chassis that sets the tolerances for all play within the robot. Try to avoid the accumulation of play within the design. This advice would apply to all robot designs except certain exceptional designs that actually rely on the flexibility of the design.

Static Mechanics

We've already spoken about topics like compression, tensile strength, hardness, flexing, and materials. The derivation of the mechanical static properties of shaped materials (like compression strength, tensile strength, flexibility, etc.) is beyond this text, but this does not mean that the design has to be done blindly. If preformed materials are used, the manufacturer should be able to specify these properties for the parts in question. If the manufacturer cannot, then consider finding another manufacturer. The parameters in question are not difficult to calculate or measure empirically, but the engineer must have the right tools and knowledge.

If the tensile strength or compression strength of a structural member must be calculated, consider finding an ME consultant to perform the work. One other option would be to find a similar part of roughly the same shape and extrapolate the parameters. Here's one example.

Suppose you want to know the compression strength of an L-shaped beam made of a specific type of plastic. If the manufacturer has already specified the compression strength of a single slab of material with the same thickness, you have enough information to make an estimate. Simply add together the compression strength of the two flat portions of the L-beam. This estimate of the compression strength for the L-beam will probably be low, but that's just fine.

Dynamic Mechanics

The field of dynamics is vast and complicated. Even without the complications of relativistic motion, the physics and math are difficult and beyond the scope of this text. However, a couple of useful tips must be passed on.

ENERGY CALCULATIONS

It's useful to be able to estimate the energy required to make parts move within the robot. The calculations required for making these estimates vary with the types of motions involved.

Consider a bicycle. How much energy does it take to accelerate a bike to a fixed speed? Let's assume the following: The bike chassis, without the wheels, has a mass of W1. Each wheel has a mass of W2 and has a radius of R. The bike will accelerate to a speed of V. The energy of a mass moving in a straight line is

$$0.5 \times m \times v^2$$

where m is the mass and v is the velocity. Notice the similarity here with Einstein's famous $E = mc^2$ formula!

Now, if the wheels were not spinning, the energy of the bike would be

$$E = 0.5 \times (W1 + W2 + W2) \times V^2$$

But the tires are indeed spinning and contain energy as well. The energy in a mass constrained to rotate about a central point is

$$E = 0.5 \times m \times r^2 \times (d\theta/dt)^2$$

where m is the mass, r is the radius of rotation, and θ is the angular position of the rotating mass. This is the best equation for measuring the energy, but there's an easier way.

If all the weight, W1, of the tire were at the edge (radius r), then each particle of the tire would be moving at a speed of V. Each tire's rotational energy would be

$$E \;=\; 0.5 \,\times\, W2 \,\times\, V^2$$

As a practical matter, not all of the tire's mass is at the rim. Some of the mass is within the spokes. For the bicycle, the previous equation is a good conservative estimate, but for wheels shaped like a hockey puck, significant weight would exist on the inside of the wheel, closer to the axle. The rotational energy of the wheel would be lower than the previous number. It would take a bit of calculus to compute the proper number. However, estimating the number can be done in an easier way. The energy of a rotating particle of mass grows as r^2, but the number of such particles grows with the circumference of travel as r increases. The calculus shows the energy increasing as r^3. If we want to estimate the rolational energy in the wheel, we want to find $r1$ such that $r1^3 = 0.5\, r^3$. This radius, r1, turns out to be about 80 percent of r. Although the outside of the wheel might be moving at a speed of V, the average part of the wheel at a radius of r1 is moving at $.8 \times V$. So a good first approximation for the rotational energy in a solid core wheel would be

$$E \;=\; 0.5 \,\times\, W2 \,\times\, (0.8 \,\times\, V)^2 \;=\; 0.32 \,\times\, W2 \,\times\, V^2$$

This would put the total energy within the bike between the following two energies:

- **High estimate**   This estimate assumes all the mass of the wheel is at the edge near the rim:

$$E \;=\; 0.5 \,\times\, (W1 \,+\, W2 \,+\, W2) \,\times\, V^2 \,+\, 2 \,\times\, (0.5 \,\times\, W2 \,\times\, V^2)$$
$$E \;=\; 0.5 \,\times\, (W1 \,+\, 4 \,\times\, W2) \,\times\, V^2$$

- **Low estimate**   This estimate assumes all the mass of the wheel is evenly distributed throughout the wheel:

$$E \;=\; 0.5 \,\times\, (W1 \,+\, W2 \,+\, W2) \,\times\, V^2 \,+\, 2 \,\times\, (0.32 \,\times\, W2 \,\times\, V^2)$$
$$E \;=\; 0.5 \,\times\, (W1 \,+\, 2.64 \,\times\, W2) \,\times\, V^2$$

Do not forget that imparting energy to parts within the robot cannot be done efficiently. These equations are only theoretical and are used to estimate only the energy within the moving parts. The energy needed to accelerate the parts to the speeds in question will be greater than the estimate.

NATURAL FREQUENCIES

We've already covered natural vibration in a previous chapter. All mechanical structures will vibrate easily at specific "natural" frequencies. The materials and the structure contribute to this particular type of vulnerability. At worst, the robot may shake apart. At best, the robot may make noise as it moves. The best way to eliminate this problem is to vary the design in ways that make cooperative vibrations less likely. Notice that the solutions for damping out vibrations are much the same as adding friction to our second-order control system.

Here are a couple web sites about natural frequency vibrations:

- www.ideers.bris.ac.uk/resistant/vibrating_build_natfreq.html
- www.newport.com/Vibration_Control/Technical_Literature/Fundamental_of_Vibration/Fundementals_of_Vibration/

HEAT TRANSFER

A couple of short notes must be made about heat transfer. Often heat must be taken out of a component. Heatsinks, for example, remove heat from integrated circuits like microprocessors. Although heat transfer is a general problem, we can use a processor and a heatsink in our example without a loss of generality. Heat flows from the processor, through the heatsink, and into the ambient air. Each component has a well-specified thermal impedance that can be used to measure its effectiveness. Low thermal impedance means the component can transfer heat more effectively. Here's how the calculations are done.

Suppose the processor dissipates 20 watts, that the ambient air is at 25 degrees Celsius, and that the thermal impedance of the heatsink is 2 degrees Celsius per watt. The processor will rise to a temperature of

$$25 + 2 \times 20 = 65 \ degrees \ Celsius$$

This temperature may be too high for the processor. If that's the case, then lower the temperature of the ambient air, get a heatsink with a lower thermal impedance, or find a lower-energy processor.

Here are a few web sites describing thermal impedance calculations:

- http://sound.westhost.com/heatsinks.htm#asample%20calc
- www.hardwarecentral.com/hardwarecentral/tutorials/743/1/
- www.hardwarecentral.com/hardwarecentral/tutorials/950/1/

INDEX

ABOUT THE AUTHOR

Charles Bergren has an MSEE from Cornell University and has been a top-notch EE for over 30 years. He has a consulting engineering firm, http://www.bdesigncorp.com, and has served as VP of Engineering in numerous companies. His design work includes:

- Robotic Tape Libraries   Building high-speed network control processors for storage systems.
- Voice Processing Systems   For voice recognition and text-to-speech applications
- Vision Systems   For the visual tracking of bacteria and DSP
- Video Systems   For MPEG video compression and satellite broadcasting
- Power Control Systems   For solar powered systems and nuclear reactor monitors
- Communication Systems   For wireless RF and free space laser networks

He has taught academic and technical courses from kindergarten through college. His work teaching and playing lacrosse has led to two observations:

- Nothing prepares you for the world like teaching proxy warfare to JHS boys
- After people chase you with sticks for 2 hours, nothing else can hurt you the rest of the week.